T0300996

Design, Development and Analysis of Lunar Crescent Visibility Criterion with Python

The analysis of lunar crescent visibility criterion is vital in providing a comparative insight for predicting the visibility and suitability for Hijri calendar determination. While there have been previous attempts to measure the performance of lunar crescent visibility criterion, these attempts only apply to a singular analysis and not a comparative examination.

Design, Development and Analysis of Lunar Crescent Visibility Criterion with Python explores the development of an analysis tool for lunar crescent visibility criterion using an integrated lunar crescent visibility database. The analysis tool, called HilalPy and HilalObs, was developed in the form of a Python library, so that it can be integrated into other software and webpages to enable deployment into various operating systems. This book will provide useful insights for the future development of lunar crescent visibility criterion, particularly for calendrical purposes.

Key Features:

- Presents an analysis tool for lunar crescent visibility based on an integrated lunar crescent report database.
- Offers researchers and practitioners the capability to perform comparative analyses of the percentage of reliability in lunar crescent visibility criteria, thereby contributing to the construction of robust criteria for determining the Hijri calendar.
- Provides a comprehensive resource for researchers, students, policymakers, and practitioners involved in the determination of the Hijri calendar.

Muhamad Syazwan Bin Faid, PhD, Department of Islamic Studies, Centre for General Studies and Co-curricular, Universiti Tun Hussein Onn, Malaysia.

Mohd Saiful Anwar Mohd Nawawi, PhD, Department of Fiqh and Usul, Academy of Islamic Studies, Universiti Malaya.

Mohd Hafiz Mohd Saadon, PhD, Department of Fiqh and Usul, Academy of Islamic Studies, Universiti Malaya.

Design, Development and Analysis of Lunar Crescent Visibility Criterion with Python

Muhamad Syazwan Bin Faid, Mohd Saiful Anwar
Mohd Nawawi and Mohd Hafiz Mohd Saadon

CRC Press
Taylor & Francis Group
Boca Raton London New York

CRC Press is an imprint of the
Taylor & Francis Group, an **informa** business

Front cover image: CRC Press

First edition published 2024
by CRC Press
2385 NW Executive Center Drive, Suite 320, Boca Raton FL 33431

and by CRC Press
4 Park Square, Milton Park, Abingdon, Oxon, OX14 4RN

CRC Press is an imprint of Taylor & Francis Group, LLC

© 2025 Muhamad Syazwan Bin Faid, Mohd Saiful Anwar Mohd Nawawi and Mohd Hafiz Mohd Saadon

Reasonable efforts have been made to publish reliable data and information, but the author and publisher cannot assume responsibility for the validity of all materials or the consequences of their use. The authors and publishers have attempted to trace the copyright holders of all material reproduced in this publication and apologize to copyright holders if permission to publish in this form has not been obtained. If any copyright material has not been acknowledged, please write and let us know so we may rectify in any future reprint.

Except as permitted under U.S. Copyright Law, no part of this book may be reprinted, reproduced, transmitted, or utilized in any form by any electronic, mechanical, or other means, now known or hereafter invented, including photocopying, microfilming, and recording, or in any information storage or retrieval system, without written permission from the publishers.

For permission to photocopy or use material electronically from this work, access www.copyright .com or contact the Copyright Clearance Center, Inc. (CCC), 222 Rosewood Drive, Danvers, MA 01923, 978-750-8400. For works that are not available on CCC please contact mpkbookspermissions @tandf.co.uk

Trademark notice: Product or corporate names may be trademarks or registered trademarks and are used only for identification and explanation without intent to infringe.

ISBN: 978-1-032-86692-5 (hbk)
ISBN: 978-1-032-88105-8 (pbk)
ISBN: 978-1-003-53619-2 (ebk)

DOI: 10.1201/9781003536192

Typeset in Times
by Deanta Global Publishing Services, Chennai, India

Contents

List of Figures

List of Tables

Acknowledgements

Alhamdulillah, praise be to Allah SWT for His Rahmah and with His permission we managed to complete this book successfully. With His permission, too, we gained the enthusiasm and spiritual strength needed in weathering the trials and tribulations on my journey. We would also like to extend our grateful thanks to the parties who have been immensely helpful and supportive throughout the time we worked on this book. Without their help, cooperation, and support, we may not have been able to come this far. Finally, to all parties directly or indirectly involved who contributed to this study, we pray that you will be blessed by Allah SWT and bestowed with the best rewards. Thanks for everything.

Issue in Lunar Crescent Visibility Criterion

1

1.1 INTRODUCTION

The new Hijri month is determined by the sighting of a lunar crescent. The practice of using the sighting of a lunar crescent originates from the hadith of Nabi Muhammad SAW, as narrated by 'Abd Allāh bin 'Umar RA, Allah's Messenger mentioned Ramaḍ ān and spoke:

.لاَ تَصُومُوا حَتَّى تَرَوُا الْهِلاَلَ، وَلاَ تُفْطِرُوا حَتَّى تَرَوْهُ، فَإِنْ غُمَّ عَلَيْكُمْ فَاقْدُرُوا لَه

> Do not fast unless you see the crescent (of Ramaḍ ān), and do not give up fasting till you see the crescent (of Shawwāl), but if the sky is overcast (if you cannot see it), then act on estimation (i.e., count Sha'bān as 30 days).[1]

According to Ibn Ḥajar al-'Asqalānī, this hadith states that the determination of a new Hijri month is through sighting of the lunar crescent on every 29th of a Hijri month. If the lunar crescent is not sighted, either due to cloudy skies or other hindrance, a Hijri month should be completed on its 30th day.[2] As Islamic territories have expanded into other countries and cover various nations, the practice of Hijri month determination has changed from lunar crescent sighting to estimating the parameters of a lunar crescent sighting on

DOI: 10.1201/9781003536192-1

the 29th of each Hijri month. This estimation of parameters facilitates yearly Hijri calendar planning for Muslims and overcomes any case of inconsistencies in lunar crescent visibility reports.[3] The estimation of a lunar crescent sighting is known as the lunar crescent visibility criterion (*kriteria imkān al-ru'yah*) among Brunei, Indonesia, Malaysia, and Singapore. In estimating the visibility of a lunar crescent, a collection of data from lunar crescent observation is analyzed[4] This ensures the reliability of a criterion.

1.2 BACKGROUND OF LUNAR CRESCENT VISIBILITY CRITERION DESIGN

The research on formulating a lunar crescent visibility criterion has been ongoing since the medieval era, pioneered by al-Khawārizmī.[5] In the late twentieth century, there was a reemergence among Muslim scholars in this area of research, probably due to the increasing number of erroneous visibility reports.[6] This was marked by the creation of a lunar crescent visibility criterion during the Istanbul Declaration 1978,[7] followed by Ilyas in 1981, Fatoohi in 1998, Ramadhan et al. in 2014, Alrefay et al. in 2018, Ahmad et al. in 2020, and most recently Utama et al. in 2023.[8]

To date, there have been at least 20 lunar crescent visibility criteria established from the ancient era, medieval era, and modern era. Each criterion is unique from the others. The lunar crescent visibility criterion formulated by Ilyas in 1981, for example, was designed as a criterion for the international lunar datelines, which is why it has a higher visibility line compared to other lunar crescent visibility criteria. The Ministers of Religion of Brunei, Indonesia, Malaysia and Singapore (MABIMS) criterion was set lower than the majority of lunar crescent visibility criteria as it confines its dataset to Brunei, Indonesia, Malaysia and Singapore, thus serving its purpose as a Hijri calendar criterion for these countries. These examples explain why the criteria are unique and distinct from one another.[9]

The unique nature of the lunar crescent visibility criterion extends to those that share the same variable. Examples are the Odeh and Qureshi lunar crescent visibility criteria; both criteria use the same variables of lunar crescent visibility, which are sun–moon altitude and lunar width; however, the parameters that apply are different. A similar vein can be observed in the lunar crescent visibility criterion formulated by Maunder and Fotheringham. Maunder's lunar crescent visibility line is lower than Fotheringham's to accommodate negative sighting,[10] whereas Fotheringham's visibility graph is higher as it favours a positive lunar crescent observation. Although Maunder added a small dataset, most of the data are the same as Fotheringham's. This demonstrates a case where the same dataset and variables can lead to differences in parameters. In fact, almost all of the lunar crescent criteria do not share the same parameters and expressions, even though they use the same variables. This shows the inconsistent nature of the lunar crescent visibility criterion. Such inconsistency has led to confusion in choosing the most reliable criterion for lunar crescent visibility prediction. As lunar crescent visibility is used to determine the Hijri month, choosing the most reliable criterion to determine lunar crescent visibility during the 29th day of the Hijri month is paramount. An incorrect prediction of lunar crescent visibility will result in confusion among Muslims and raise doubt towards the authenticity of the commencement of the new Hijri month. This calls for a comparative assessment of the various lunar crescent visibility criteria. A comparative assessment can determine which criteria are the most suitable for lunar crescent visibility prediction, and which are not effective as a reference for Hijri calendrical purposes.

1.3 ISSUE IN ANALYZING THE LUNAR CRESCENT VISIBILITY CRITERION

Three problem statements have been identified with regard to the lunar crescent visibility criterion assessment. The explanation of the highlighted problem statements is discussed in the below discussions.

1.3.1 Variety of Astrometry Libraries Used for Calculation in Lunar Crescent Visibility Report

The issue of inconsistency of lunar crescent visibility criteria can be explained by the usage of different astrometry libraries when calculating the lunar crescent visibility report. A different astrometry library would result in different values of parameter for the lunar crescent visibility criterion. Examples of differences in the calculated values using different astrometry libraries are demonstrated in Table 1.1. The data is calculated using the PyEphem library,[11] which applies a high-accuracy ephemeris used for various applications such as the construction of satellite telemetry,[12] weather monitoring research,[13] and the

TABLE 1.1   Recalculated Data of Lunar Crescent Visibility Report

NO	DATA	LUNAR POSITION ORIGINAL REPORT	RECALCULATED REPORT	ORIGINAL SOURCE ASTROMETRY LIBRARY
1	Hasanzadeh (2012)	Elongation: 5°42'	6°41	AH Sultan (2004)
2	Sultan (2007)	Elongation: 5°30'	4°59'	AH Sultan (2004)
3	Salimi (2005)	Altitude: 2°00'	2°12'	Almanak Hisab Rukyat
4	Djamaluddin (2001)	Altitude: 3°36'	2°24'	Djamaluddin (2001)
5	JAKIM (2012)	Elongation: 5°18'	5°33'	Almanak Falak
6	Qureshi (2010)	Altitude: 5°24'	3°19'	Qureshi (2005)

Source: Researcher Data.

development of robotic telescopes.[14] Its accuracy and utility in various fields have proven the PyEphem library's reliability as an astrometry library for lunar crescent visibility report.

A pilot investigation found that there are at least six data points of lunar crescent visibility reports that contradict each other when recalculated using the PyEphem library, with a maximum contradiction of 2°05'. These data are reference data used to construct the parameters of the lunar crescent visibility criterion. Another issue stemming from various preferences of astrometry libraries is the variety in the calculated dates of the Hijri months obtained. Some data of lunar crescent visibility are located at the borderline of the lunar crescent visibility criterion, either visible or not visible, according to a given criterion. This borderline data, if recalculated using a different astrometry library, will result in a different determination of a new Hijri month.

This serves as evidence that different astrometry libraries used will result in different parameters for the lunar crescent visibility criterion. Thus, the first problem statement is: How do we solve the issue of using different astrometry libraries for calculations in lunar crescent visibility reports?

1.3.2 Data Locality of Lunar Crescent Visibility Reports

A scientific result can be applicable to any given location, time, and condition, with certain limitations.[15] Thus, the experimental data used in a scientific study must represent various testing environments, including location, time, and condition in accordance with the relevant scientific objectives.[16] As a product of a scientific result, the lunar crescent visibility criterion must be applicable to any given location, time, and condition in accordance with its purpose. The lunar crescent visibility criterion must also be experimented with using data that represent various locations, times, and conditions. If a lunar crescent visibility criterion is experimented in a limited environment, then the results will only apply to its confined environment and possibly falter when tested outside its environment.[17]

Some researchers in the area of lunar crescent visibility criterion use specific locality databases in constructing their criterion, while there are also studies that include various localities in their research such as Odeh and Schaefer.[18] Ahmad et al. from Malaysia primarily used lunar crescent visibility reports from Malaysia for their lunar crescent criterion.[19] Ramadhan et al. from Indonesia used lunar crescent visibility reports from Indonesia.[20] The works of Alrefay et al. and Hoffman also fall under the same vein.[21] This makes their lunar crescent visibility criteria theoretically applicable only to their own limited data localities.

To assess and compare the accuracy of any lunar crescent criterion to another, data of the lunar crescent must be based on various localities, testing conditions, and methods of sighting. This raises the question of how to solve the issue of data collection of lunar crescents that have various localities and testing conditions.

1.3.3 The Absence of Confirmation Methodology for Lunar Crescent Visibility

Currently, although the majority of lunar crescent sighting practitioners use digital imaging to validate their sightings, there are still cases where the visibility of the lunar crescent is falsely reported without any supporting digital proof. Table 1.2 contains false-positive data on lunar crescent sightings from Malaysia and Indonesia. The records of lunar crescent sightings in Table 1.2 are found to be observed by a sizable number of witnesses, and it is impossible to dismiss their data. For this reason, data in Table 1.2 are accepted as a base

TABLE 1.2   Selected Data of Arguable Lunar Crescent Visibility

LOCATION	DATE	MOON ALTITUDE	ELONGATION	MOON AGE	LIMITING NAKED-EYE PARAMETER
Melaka, Malaysia	11 June 1983	3°46′	3°01′	6 hours 8 minutes	Altitude = 4.0°
Pelabuhan Ratu, Indonesia	11 June 1983	1°52′	1°55′	6 hours 24 minutes	elongation = 7.5° age = 10 hours
Banda Acheh, Indonesia	25 April 1990	1°26′	6°02′	7 hours 28 minutes	

for the MABIMS lunar crescent visibility criterion in determining the new Hijri month in Malaysia, Indonesia, Brunei, and Singapore.[22] However, some of the data in Table 1.2 fall well below the visible threshold parameter when compared to other studies of the lunar crescent sighting, such as Fatoohi.[23]

Other regions of the world have reported false-positive lunar crescent sightings as well. In Saudi Arabia, Kordi discovered that at least 24 falsely positive reports of sightings of the lunar crescent were accepted.[24]n Nigeria in November 2017, there were cases of false-positive lunar crescent sightings.[25] Hence, there is a need for a lunar crescent sighting report confirmation methodology. In the past, Ramadhan et al. outlined a reduction parameter to confirm sightings of the lunar crescent.[26] Their confirmation methodology is as follows:

- If the arc of vision is smaller than 4 degrees, an accepted observation must be conducted by three or more independent groups. Additionally, at least one report needs to be cross-checked with astronomical calculations.
- If the recorded time of observation was later than the anticipated time of moonset, the data was rejected.
- Observation of bright background objects such as Venus or Mercury is rejected as it could confuse the observer.

The parameter described appears to be accurate, although it is not a perfect confirmation tool. It is discovered that the last three lunar crescent sightings that pass the Djamaluddin methodology are too close to the horizon and impossible to be sighted by the naked eye, as these data were recorded by three different groups and included a thorough observation report, which Djamaluddin's methodology considered acceptable. Schaefer developed an algorithm to determine lunar crescent visibility. His algorithm considers sky brightness, lunar brightness, and human-detectable contrast threshold. Through extensive projects of Moonwatch,[27] a nationwide project to observe and investigate the limit of lunar crescent visibility in the United States, the algorithm's dependability is tested. However, it has been noted that Schaefer's method is inconsistent because Schaefer changed the variable in the algorithm each time it was published. However, Schaefer finally published the final edition of his algorithm in 2000. Schaefer's algorithm, which considers various effects of lunar crescent sightings, including atmospheric extinction, light pollution, and human eye sensitivity, is effective and should be replicated for future use.

A new confirmation process for the lunar crescent visibility report is required due to a flaw in the current confirmation of lunar crescent visibility reports. The methods must consider the parameters for spotting the lunar crescent, which include the brightness of the moon, the sky, light pollution, atmospheric extinction, and the sensitivity of human eyes. The method described in this study is intended to serve as a confirmation tool for the reported sightings of the lunar crescent. The method considers astrophysical elements such as atmospheric extinction, light pollution, sky and lunar brightness, and human contrast threshold when determining the visibility of the lunar crescent.

1.3.4 The Absence of Analysis Tools for Lunar Crescent Visibility Criterion

To date, there have been limited attempts to assess the accuracy and reliability of lunar crescent visibility criteria. One such example is Fatoohi, in his thesis, where he examined various lunar crescent visibility criteria such as those formulated by Danjon, Yallop, Bruin, and Ilyas.[28] Fatoohi's data included lunar crescent observations from the Babylonian era, the medieval era, and Schaefer's data of lunar crescent visibility. However, Fatoohi's research was conducted 20 years ago, making his assessment outdated.

Hoffman produced a rational design to assess the reliability of a lunar crescent visibility criterion. Hoffman's rational design is based on a linear relationship between two lunar crescent visibility parameters. The problem with Hoffman's rational design is that his work is not replicable, given that it has several uncertain coefficients, such as k, v, and x. In addition to being difficult to formulate, the coefficients will result in different formulations being obtained by different researchers, thus making any endeavour of comparative assessment difficult.[29] Therefore, there is a need for a comparative assessment tool to compare the different criteria that are replicable and capable of producing consistent results, even when used by multiple researchers. Combining these research problems, this book entitled *Design, Development and Analysis of Lunar Crescent Visibility Criterion* is published.

NOTES

1. Abū ʻAbd Allāh Muḥammad bin Ismāʻīl al-Bukhārī, *Al-Jāmiʻ al-Ṣaḥ ī ḥ* (Kaherah: al-Matbaʻah al-Salafiyyah, 1983), (Kitāb al-Ṣawm, Bāb Qawl al-Nabī SAW: Idhā Raʼaytum al-Hilāl fa Ṣ ūmū, wa Idhā Raʼaytumūhu fa Afṭirū, no. hadith: 1909).
2. Ibn Ḥajar al-ʻAsqalānī, *Fath al-Bārī bi Sharḥ Ṣaḥ ī ḥ al-Bukhārī: Wa Maʼahu Tawjīh al-Qārī ilā al-Qawāʼid wa Fawāʼid al-Uṣ ūliyyah wa al-Ḥadīthiyyah wa al-Isnādiyyah*, ed. Abd al-ʻAziz ibn ʻAbd Allah Ibn Baz (Beirut: Dār al-Fikr, 1993).
3. Yūsuf al-Qaraḍ āwī, *Taysīr al-Fiqh fī Ḍaw' al-Qur'ān wa al-Sunnah: Fiqh al-Ṣiyām: Islamic Jurisprudence of Fasting* (Beirut: Mawsūʻah al-Risālah, 1994); and Hamza Yusuf, *Caesarean Moon Births: Calculations, Moon Sighting and the Prophetic Way* (Berkeley: Zaytuna Institute, 2010).
4. Mohd Saiful Anwar Mohd Nawawi et al., "Relevensi Penggunaan Kriteria Imkanurrukyah Dalam Penentuan Awal Bulan Ramadan Dan Syawal Di Malaysia [Relevance of the Use of Imkanurrukyah Criteria in Determining the Early Month of Ramadan and Syawal in Malaysia]," *Jurnal Falak* 1 (2015): 99; Mohd Saiful et al., "Application of Scientific Approach to Determine Lunar Crescent' s Visibility," *Middle -East Journal of Scientific Research* 12, no. 1 (2012): 96–100, https://doi.org/10.5829/idosi.mejsr.2012.12.1.1672.
5. Jan P. Hogendijk, "Three Islamic Lunar Visibility Calendar," *Journal of History of Astronomy*, 1988; David A. King, *Astronomy in the Service of Islam* (London: Routledge, 1993), *19*(1), 29–44.
6. M. Ilyas, *Sistem Kalendar Islam Dari Perspektif Astronomi [The Islamic Calendar System from an Astronomical Perspective]* (Kuala Lumpur: Dewan Bahasa dan Pustaka, 1997).
7. Susiknan Azhari, "Cabaran Kalendar Islam Global Di Era Revolusi Industri 4.0," *Jurnal Fiqh* 18, no. 1 (June 24, 2021): 117–34, https://doi.org/10.22452/fiqh .vol18no1.4.
8. M. Ilyas, "Lowest Limit of W in the New Moon's Visibility," *Quarterly Journal of Royal Astronomy Society*, 1981, 22, 154–59; Louay Fatoohi, "First Visibility of the Lunar Crescent and Other Problems in Historical Astronomy" (Durham University, 1998); T. B. Ramadhan, Thomas Djamaluddin, and J. A. Utama, "Reevaluation of Hilaal Visibility Criteria In Indonesia by Using Indonesia and International Observational Data," in *Proceeding of International Conference on Research, Implementation and Education of Mathematics and Sciences* (Yogjakarta, 2014); T. Alrefay et al., "Analysis of Observations of Earliest Visibility of the Lunar Crescent," *The Observatory* 138, no. 1267 (2018), 267–91; Nazhatulshima Ahmad et al., "A New Crescent Moon Visibility Criteria Using

Circular Regression Model: A Case Study of Teluk Kemang, Malaysia," *Sains Malaysiana* 49, no. 4 (2020): 859–70, https://doi.org/10.17576/jsm-2020-4904 -15; Nazhatulshima Ahmad et al., "Analysis Data of the 22 Years of Observations on the Young Crescent Moon at Telok Kemang Observatory in Relation to the Imkanur Rukyah Criteria 1995," *Sains Malaysiana* 51, no. 10 (October 31, 2022): 3415–22, https://doi.org/10.17576/jsm-2022-5110-24; J. A. Utama et al., "Young Lunar Crescent Detection Based on Video Data with Computer Vision Techniques," *Astronomy and Computing*, June 2023, 100731, https://doi.org/10 .1016/j.ascom.2023.100731.

9. Abdul Mufid and Thomas Djamaluddin, "The Implementation of New Minister of Religion of Brunei, Indonesia, Malaysia, and Singapore Criteria towards the Hijri Calendar Unification," *HTS Teologiese Studies/Theological Studies* 79, no. 1 (June 30, 2023), https://doi.org/10.4102/hts.v79i1.8774; M. Ilyas, "Lunar Calendars: The Missing DateLines," *The Journal of the Royal Astronomical Society of Canada* 80 (1986): 328–35.

10. E. Walter Maunder, "On the Smallest Visible Phase of the Moon," *The Journal of the British Astronomical Association* 21 (1911): 355–62; J. K. Fotheringham, "On The Smallest Visible Phase of the Moon," *Monthly Notices of the Royal Astronomical Society* 70 (1910): 527.

11. B. C. Rhodes, "PyEphem: Astronomical Ephemeris for Python," Astrophysics Source Code Library, record ascl:1112.014, 2011.

12. Daniel J. White et al., "SatNOGS: Satellite Networked Open Ground Station," *Engineering Faculty Publications*, 2015.

13. Daniel S. Krahenbuhl et al., "Monthly Lunar Declination Extremes' Influence on Tropospheric Circulation Patterns," *Journal of Geophysical Research Atmospheres*, 2011, 116, D23121, 1–5. https://doi.org/10.1029/2011JD016598.

14. Mauro Stefanon, "The Software for the Robotization of the TROBAR Telescope," *Advances in Astronomy*, 2010, 1–5, https://doi.org/10.1155/2010/785959.

15. Norwood Russell Hanson, "Is There a Logic of Scientific Discovery?," *Australasian Journal of Philosophy* 38, no. 2 (1960): 91–106.

16. Karl Popper, *The Logic of Scientific Discovery* (New York and London: Routledge, 2005).

17. Herbert A. Simon, "Does Scientific Discovery Have a Logic?," *Philosophy of Science* 40, no. 4 (1973): 471–80.

18. Mohammad Odeh, "New Criterion for Lunar Crescent Visibility," *Experimental Astronomy* 18 (2004): 39–64; Bradley Schaefer, "Lunar Crescent Visibility," *Icarus* 107 (1994): 1.

19. Nazhatulshima Ahmad et al., "A New Crescent Moon Visibility Criteria Using Circular Regression Model: A Case Study of Teluk Kemang, Malaysia," *Sains Malaysiana* 49, no. 4 (2020): 859–70; Nazhatulshima Ahmad et al., "Analysis Data of the 22 Years of Observations on the Young Crescent Moon at Telok Kemang Observatory in Relation to the Imkanur Rukyah Criteria 1995," *Sains Malaysiana* 51, no. 10 (October 31, 2022): 3415–22.

20. T. B. Ramadhan, Thomas Djamaluddin, and J. A. Utama, "Reevaluation of Hilaal Visibility Criteria In Indonesia by Using Indonesia and International Observational Data," in *Proceeding of International Conference on Research, Implementation and Education of Mathematics and Sciences* (Yogjakarta, 2014).

21. T. Alrefay et al., "Analysis of Observations of Earliest Visibility of the Lunar Crescent," *The Observatory* 138, no. 1267 (2018); Roy E. Hoffman, "Observing the New Moon," *Monthly Notices of the Royal Astronomical Society* 340, no. 3 (April 11, 2003): 1039–51, https://doi.org/10.1046/j.1365-8711.2003.06382.x.
22. Mohd Saiful Anwar Mohd Nawawi et al., "Sejarah Kriteria Kenampakan Anak Bulan di Malaysia," *Journal of Al-Tamaddun* 10, no. 2 (2015/12/31), https://doi .org/10.22452/JAT.vol10no2.5, https://ejournal.um.edu.my/index.php/JAT/article /view/8690.
23. Louay Fatoohi, "First Visibility of the Lunar Crescent and Other Problems in Historical Astronomy" (PhD, University of Durham, 1998).
24. Ayman Kordi, "The Psychological Effect On Sighting of The New Moon," *Observatory* (2003): 219–23.
25. Abdulmajeed Bolade Hassan-Bello, "Sharia and Moon Sighting and Calculation Examining Moon Sighting Controversy in Nigeria," *Al-Ahkam* 30, no. 2 (2020): 215–52.
26. T. B. Ramadhan, Thomas Djamaluddin, and J. A. Utama, "Reevaluation of Hilaal Visibility Criteria In Indonesia by Using Indonesia and International Observational Data" (paper presented at the Proceeding of International Conference On Research, Implementation And Education of Mathematics and Sciences, Yogjakarta, 2014).
27. L. Doggett, Bradley Schaefer, and L. Doggett, "Lunar Crescent Visibility," *Icarus* 107, no. 2 (1994), https://doi.org/10.1006/icar.1994.1031.
28. Louay Fatoohi, "First Visibility of the Lunar Crescent and Other Problems in Historical Astronomy" (PhD, University of Durham, 1998); Andre-Loius Danjon, "Le Croissant Lunaire [The Lunar Crescent]," *L'Astronomie: Bulletin de La Société Astronomique de France*, 1936, 57–65; B. D. Yallop, *A Method for Predicting the First Sighting of the Crescent Moon, Nautical Almanac Office* (Cambridge: Nautical Almanac Office, 1998); Frans Bruin, "The First Visibility of the Lunar Crescent," *Vistas in Astronomy* 21 (January 1977): 331–58, https:// doi.org/10.1016/0083-6656(77)90021-6; M. Ilyas, "Lunar Crescent Visibility Criterion and Islamic Calendar," *Quarterly Journal of Royal Astronomical Society* 35 (1994): 425–61.
29. Roy E. Hoffman, "Observing the New Moon," *Monthly Notices of the Royal Astronomical Society* 340, no. 3 (April 11, 2003): 1039–51; Roy E. Hoffman, "Rational Design of Lunar-Visibility Criteria," *The Observatory* 125 (2005): 156–68; Roy E. Hoffman, "Back to Back Crescent Moon," *The Observatory* 129, no. 1208 (2009): 1–5.

Design and Development of Lunar Crescent Visibility Criterion

2

2.1 INTRODUCTION

This chapter discusses a review on the lunar crescent visibility criterion. The chapter embarks with a study on the parameter of lunar crescent visibility, which are lag time, moon age, arc of light, arc of vision, difference in azimuth, and lunar crescent width. The study includes a brief history of the parameter, its application in determining the lunar crescent visibility and its sighting threshold. Then, the chapter followed by the study of lunar crescent visibility criterion. The study includes lunar crescent visibility criterion from ancient era, such as Babylonian and Indian lunar crescent visibility criterion, followed by lunar crescent visibility criterion from Middle Age, and lastly modern lunar crescent visibility criterion. Each of the studies includes parameter and design of a criterion, its strengths, and weaknesses. This chapter is then concluded by a study on the assessment of lunar crescent visibility criterion.

DOI: 10.1201/9781003536192-2

2.2 LUNAR CRESCENT VISIBILITY CRITERION

Predicting the lunar crescent visibility is an issue that has attracted interest throughout history. The oldest records of lunar crescent visibility date more than 2,000 years ago in the Babylonian age.[1] Prediction of a lunar crescent visibility gathers the attention of Muslim scholar during the Middle Age as Muslims require prediction of lunar crescent sighting to determine their new Hijri month. Likewise, the Jewish, Babylonian, and Indian civilization also found interest in lunar crescent visibility for the purpose of calendrical computation.[2]

Muslim interest in determining the visibility of the lunar crescent continues into the present date. Determining the visibility of the lunar crescent is regarded as the utmost important by Muslims as it is directly tied with the celebration of Ramadhan, Eid Fitri, and Eid Adha. As Muslim population in 2030 is expected to be around 2.2 billion,[3] the determination of limiting visibility of the lunar crescent is likely to be one of the non-trivial research in astronomy that has the greatest impact in the modern world.[4]

Defining the limits of the lunar crescent visibility is a complicated matter. Examining a lunar crescent visibility requires study on human eye detection, telescopic limiting magnitude, brightness of the twilight, brightness of the lunar crescent, effect of the atmosphere and air mass, light pollution, and lunar width. Lunar crescent visibility also requires contemplation of human psychology and cloud condition, which cannot be fully interpreted in a quantified manner, even in modern astronomy.

2.3 THRESHOLD OF VISIBILITY

The parameters of a lunar crescent visibility can be classified into lag time, moon age, arc of vision, arc of light, different in azimuth, and width. These parameters function as limits in determining lunar crescent visibility. If a lunar crescent position passes a certain parameter threshold, then the lunar crescent would have a higher chance of visibility.

2.3.1 Lag Time

Lag time is defined as the time difference in minutes between sunset and moonset for evening observation. Lag of time can also be defined as arc of separation. The lag time is first introduced by Babylon,[5] and then Hindu civilization.[6] Lag time is primarily used as a supplement parameter to increase the probability of lunar crescent sighting.[7] Scholars that still use lag time are Caldwell and Gautschy.[8] The advantage of Lag time' is that it provides an easy rule to determine lunar crescent visibility and reject unverified reports.[9] However, its main drawback is that it is heavily dependent on latitude. The world record of lag time visibility for the naked eye is 29 minutes, 20 minutes for optical-aided observation, and 20 minutes for telescopic observation.

2.3.2 Moon Age

Moon age is defined as the time in hours of the moon from its conjunction sunset. Moon age was one of the oldest parameters to determine the lunar crescent sighting. One of the first modern works on the moon age parameter of lunar crescent visibility was by Fotheringham in 1921.[10] Fotheringham argues that an hour increment of moon age increases the probability of visibility by 2.7 times. Moon age weaknesses are that moon age is influenced by solar longitude and the latitude of the location. The moon age parameter is also only suitable for a latitude below 30 degree and not for higher latitude.[11] Its strength of moon age parameter is it is able to provide a simplified rule to validate a lunar crescent visibility report. The world record of moon age parameter for lunar crescent visibility is 15 hours and 1 minute for naked eye, 12 hours and 23 minutes for optical-aided observation, 14 hours and 9 minutes for telescopic observation and 54 minutes for CCD observation.

2.3.3 Arc of Light

Arc of light or elongation is defined as the angle between the lunar crescent and the sun. Arc of light was introduced by Ibnu Yunus in eighth century during the Middle Age,[12] and was reintroduced again in the modern era by Danjon in 1936.[13] Danjon described that elongation contributes to the shortening of a lunar crescent. Danjon argued that the length of lunar crescent is directly proportional to elongation, until it reaches its limit at 7 degrees of elongation. Elongation is one of the most reliable parameters for lunar crescent visibility and was demonstrated by Ahmad as the highest correlated parameter for visibility.[14] Elongation is extensively used by astronomers to determine the limits of visibility such as Ilyas and Caldwell.[15] The world records for elongation are 7.7 degrees at naked eye, 6.0 degrees at optical aid, 6.8 degrees at telescopic observation and 3.42 degrees at CCD imaging.

2.3.4 Arc of Vision

Arc of vision or altitude is defined as the angle from midpoint of the sun to midpoint of the moon that is perpendicular to horizon. Arc of vision as can simply be termed as the differences of altitude between the sun and moon. The arc of vision is first introduced by Fotheringham.[16] Arc of vision is widely used by astronomers as it represents the brightness of the observed lunar crescent. Higher value of arc of vision translate to lower atmospheric extinction and thus a higher level of luminance and chance of visibility. Arc of vision is, however, argues by Schaefer to have the chance of miss prediction of one out of four times.[17] Arc of vision is used by Ilyas, Nawawi, Fotheringham, Maunder, and Krauss in their lunar crescent visibility criterion.[18] The world records for altitude are 4.06 degrees at naked eye, 6.48 degrees at optical aid, 4.81 degrees at telescopic observation and 4.62 degree at CCD imaging.

2.3.5 Difference in Azimuth

Difference in azimuth, as the name suggest, is the difference of sun azimuth against moon azimuth. The difference in azimuth is introduced by Fotheringham. Bruin argues that the higher the value of difference in azimuth, the higher the chance of visibility. This parameter is used by Maunder, and Ilyas in their lunar crescent visibility criterion. The world records for difference in azimuth are 0.01 degrees for naked eye, 0.05 degrees for optical aid, 0.01 degrees for telescopic observation and 0.03 degrees for CCD imaging.

2.3.6 Lunar Crescent Width

Lunar crescent width is an angle of the illuminated area of a lunar crescent. Bruin is the first to design a lunar crescent visibility criterion using the parameter of width. Argument behind the adoption of width for a lunar crescent visibility criterion is that width represents the limiting detection size of lunar crescent by human. A higher size of a lunar crescent means a wider detectability area and thus a higher chance of successful sighting. Bruin adaptation of lunar crescent width in his visibility criterion is then followed by Yallop, Odeh, and Qureshi.[19] The world records for width are 6.45 minutes at naked eye, 5.83 degrees at optical aid.

2.4 LUNAR CRESCENT VISIBILITY CRITERION

2.4.1 Ancient Lunar Crescent Visibility Criteria

Ancient lunar crescent visibility criteria are criteria that were produced during the ancient times, which is around the period of 500 AD to 2000 BC This period observes at least two civilizations that adopted moon phase for their calendrical purpose, which are Babylonian civilization, spanning around 1894 BC to 1000 AD, and Indian civilization, spanning around 500 AD to 700 BC. Their planetary theories are somewhat different from modern planetary theory; therefore, they use a different parameter and system to determine their lunar crescent visibility criteria.

2.4.1.1 Babylonian Lunar Crescent Visibility Criterion

Most of the references on ancient lunar crescent visibility criterion refer to works of Bruin; however, Fatoohi et al. claimed that Bruin did not correctly portray the Babylonian lunar crescent visibility criterion. Bruin claimed that Babylon uses simple moon age and moon lag parameters, while the Indian uses more complicated lunar width in their lunar crescent visibility criterion. Fatoohi, on the other hand, argues that Bruin did not cite any credible source for his portrayals of Babylonian and Indian lunar crescent visibility criteria. Fatoohi then states that Babylonian lunar crescent visibility criteria are more complex, involving sun–moon velocity and trajectory. Gautschy argues that Babylon has another criterion, using 10-degree lag time as a rule, indicating a coexistence between two criteria.[20]

Fatoohi, in examining the reliability of Babylonian lunar crescent visibility criterion, using Babylonian latitude 32.6 degrees and the value of constant L+S to be 23 degrees, found that only 1.3 per cent of the positive observation is below the value of constant 23 degrees. However, in the upper value of constant, or the predicted visibility line, is only able to achieve the reliability of 69 per cent, misjudging 31 per cent of the negative observation. When criterion evaluated on various latitudes, it is found out that the criterion able to maintain its performance of predicting visibility, with only an 8 per cent positive contradiction rate; however, it unable to predict the invisibility of lunar crescent, with a 45.2 per cent contradiction rate.

Babylonian criterion, provide an introductory path for next generation astronomers in building their lunar crescent visibility criteria. Their lag time parameter found its application in Indian lunar crescent visibility criterion, al-Khawarizmi lunar crescent visibility criterion, majority of Middle Age lunar crescent visibility work, and today works on modern lunar crescent visibility criterion.

2.4.1.2 Indian Lunar Crescent Visibility Criterion

There is not much known about the Indian lunar crescent visibility criterion. The purpose of the criterion, its lunar theory and the nature of the criterion are not properly documented and extensively researched. Bruin has referred to the Indian criterion as a complex criterion that involves the calculation of lunar width. King demonstrated that Indian adapt simple 12-degree arc of separation or 48-minute lag time as its criterion, but the theory behind the computation is more complex. Indian derived the computation of arc of separation using ecliptic coordinate of the sun and moon.[21]

The performance of 48-minute lag time in determining the visibility of the lunar crescent is well documented by astronomers. The parameter is found to be dependent on latitude and contain a high contradiction rate at higher latitude particularly during Equinox and Solstice. The parameter, however, provides a simple rule to determine the lunar crescent visibility. Indian computation of crescent visibility latter influences the Middle Age scholar in computing their visibility table, particularly the adaptation of ecliptic coordinate in calculation.

2.4.2 Middle Age Lunar Crescent Criteria

Middle Age lunar crescent visibility criteria are criteria that were found during the period of 500 AD to 1901 AD. The study of lunar crescent visibility is heavily influenced by the work of Indian, Parsi, Babylonian and Greek literature of astronomy. Al-Arjabar, Zij al-Arkand, Zij al-Sindhind, Zij ash-Shah, and Ptolemy's al-Magest is one of the few translated astronomy literatures that set the background of Middle Age astronomy.[22] Following the works of their astronomical predecessor, the majority of Middle Age astronomers compiled their lunar crescent visibility criteria in the form of handbook or zijes.[23] Most of them also follow the similar planetary theory, which is ecliptic-based astronomical computation.

Where $\Delta\lambda$ is the difference between lunar and solar longitude, μ is variable that dependent on the location latitude, β is lunar latitude, and $f(n)$ is the series of the longitude limit, from the zodiacal longitude. Most of the work on lunar crescent visibility in the Middle Ages is theoretical, thus explaining the low level of their accuracy.

2.4.2.1 Al-Khwarizmi's Lunar Crescent Visibility Criterion

Al-Khawarizmi, a famous astronomer from Baghdad born in 780, recorded a lunar crescent visibility criterion in a part of a handbook. His table of lunar crescent visibility criterion is called *al-ru'yah lil Khawarizmi*, meaning the visibility according to al-Khawarizmi.[24] Al-Khawarizmi uses two main parameters in this criterion: zodiacal sign, referring to the value of longitude, and the difference of longitude between sun and moon, $\Delta\lambda$. The lunar crescent is visible when its difference in sun–moon longitude exceeds the value of $\Delta\lambda$. The value of $\Delta\lambda$ can be interpreted as an arc of separation or lag time between sunset and moonset. This indicates Indian criterion influence on al-Khawarizmi's lunar crescent visibility criterion. Fatoohi, when tested al-Khawarizmi lunar crescent visibility criterion on 33 degree latitude, found out that al-Khawarizmi criterion has contradiction rate of 31.6 per cent in predicting invisibility of lunar crescent and contradiction rate of 9.6 per cent in predicting the visibility of lunar crescent.

The contradiction rate rises significantly when evaluated with other degrees of latitude, demonstrating the unreliability of the criterion in predicting the visibility of lunar crescent. Al-Khwarizmi's criterion, despite its unreliability, indicating a relationship between the position of the earth with respect to the sun and the lag time in determining visibility of lunar crescent, can be constructed, and was the first documented work on the lunar crescent visibility criterion during the Middle Ages, which influenced its Muslim astronomy scholar in later age.

2.4.2.2 Ya'qub Ibn Tariq Lunar Crescent Visibility Criterion

Ya'qūb ibn Ṭāriq was an eighth-century Persian astronomer and mathematician who lived in Baghdad. Ibn Tariq provides a unique approach in his lunar crescent visibility criterion. While the majority of the Middle Age lunar crescent visibility criteria follow the Indian arc of separation condition in their lunar crescent visibility table, Ya'qub Ibn Tariq has produced a lunar crescent visibility use elongation parameter.[25]

Fatoohi, in examining Ibn Tariq's lunar crescent visibility criterion, has found an astonishing 4.6 per cent contradiction rate in predicting the negative and positive lunar crescent visibility for latitudes of 30–35 degrees, while in all latitudes, it yields contradiction rate of 28.6 per cent for negative observation and 6.9 per cent for positive observation. This makes Ibn Tariq lunar crescent visibility criterion the one of the most reliable criteria among Middle Age criteria. Ibn Tariq's lunar crescent visibility criterion is among the first to document work on lunar crescent criteria that introduce the inclusion of elongation in its parameter. Elongation parameter is then used by Ibn Qurra, a century later and popularized by Danjon in 1936, nine centuries later.

2.4.2.3 Thabit Ibn Qurra Lunar Crescent Visibility Criterion

Thabit Bin Qurra is Baghdad astronomer, active during the ninth century during the period of Abbasid Caliphate. He is one of the most influential individuals in understanding and reforming the original geocentric model of the universe.[26] Aligning with his geocentric model, Thabit Ibn Qurra does not follow that footstep of al-Khawarizmi, using ecliptic longitude and latitude for lunar crescent visibility determination. Ibn Qurra instead uses elongation, solar depression, and difference in azimuth for his lunar crescent visibility criterion.

Fatoohi found out that the Ibn Qurra criterion has incorrectly predicted at least 20 per cent of negative observation, indicating it is a weak criterion. The Ibn Qurra criterion, despite being low level of reliability, has shown a progressive evolution from its predecessor by using a horizontal coordinate. Horizontal coordinate enables a more direct measurement of lunar crescent parameter, in contrast to elliptical coordinates for sun and moon positions.

2.4.2.4 Al-Qallas Lunar Crescent Visibility Criterion

Maslama Ibn Ahmad al-Majriti, nicknamed al-Qallas, a Cordova astronomer in the tenth century, had produced a lunar crescent visibility table, attributed to his nickname, al-Qallas. The lunar crescent visibility table is revision of Khwarizmi table of lunar crescent visibility, with the division of the visibility function on every zodiacal sign into three, meaning al-Qallas assigned a visibility function at every 10-degree solar longitude.[27] The division could suggest more thorough research is done to define the limiting visibility of the lunar crescent. Al-Qallas lunar crescent visibility table is based on latitude of 40.87 degrees to 41.35 degrees with obliquity of 23.55 degrees to 24.00 degrees.

Similarly, with al-Khawarizmi, the lunar crescent is predicted to be visible when it exceeds the value of the arc of separation. Fatoohi, in his accuracy test of the Al-Qallas visibility table, has found out that it has a contradiction rate of 54.5 per cent in predicting negative lunar crescent reports and contradiction rate of 4.2 per cent in predicting positive lunar crescent reports. Al-Qallas revision of al-Khawarizmi is found to be less accurate than the original al-Khawarizmi lunar crescent visibility table.

2.4.2.5 Ibn Yunus Lunar Crescent Visibility Criterion

Ibnu Yunus is a celebrated astronomer from Cairo around 1000 AD. He is among the first Muslim scientists to suggest more than two parameters for lunar crescent visibility criterion. Ibnu Yunus incorporates three parameters in determining lunar crescent visibility. The first is sun–moon angular distance, or elongation, which he interpreted as lunar crescent width.[28] The second parameter is lag time, and the third is earth–moon distance, which interpreted as brightness of the lunar crescent. Ibnu Yunus lunar crescent visibility criterion designed in a condition form, similarly with Ibn Tariq lunar crescent visibility criterion. Ibn Yunus conditional style of visibility table is difficult to be comparatively assessed; hence, research that conducted Ibn Yunus criterion reliability in predicting lunar crescent visibility is not yet produced.

2.4.2.6 Al-Khāzinī Lunar Crescent Visibility Criterion

Abu al-Fath Abd Al-Rahman al-Khazini is a celebrated astronomer who lived in the Merv, currently known as Mary in Turkmenistan, during the period of 1081–1131. Al-Khazini was under the patronage of Sultan Sanjar bin Malikshah. Al-Khazini dedicated an astronomy handbook, titled *Al-Zij al-Mu'tabar al-Sanjari*, to his patron, Sultan Sanjar. His work encompasses a result of 35 years of observation. The work of al-Khazini on lunar crescent visibility table is one of the most comprehensive studies on lunar crescent visibility during his time.[29]

Al-Khazini's works on lunar crescent visibility criterion are heavily influenced by Thabit Ibn Qurra criterion, which explains why one of Khazini's lunar crescent visibility tables follows the same methodology of Thabit Ibn Qurra. Al-Khazini uses the parameter of elongation, solar depression, lag time, sun–moon longitude difference, moon angular speed and lunar altitude. Al-Khazini has two known lunar crescent visibility criteria; the first one is similar to Thabit Ibn Qurra lunar crescent visibility, using the conditional criterion framework.[30] The first Khazini lunar crescent visibility criterion is based on the lunar position in respect to earth, with parameter of elongation, lag time, solar depression, and sun–moon longitude. Al-Khazini then implies that lag time should be the deciding parameter while the other parameters are supplements to the criterion. Another al-Khazini lunar crescent visibility criterion is in the form of table. The table is categorized into three types of lunar crescent observation: General or easily visible, moderate or moderately visible, and acute or rarely visible.

Al-Khazini lunar crescent visibility table is separated into three types of observations to note the rate of successful sighting in a lunar crescent observation. Al-Khazini's second lunar crescent visibility criterion was found to be similar to the al-Khawarizmi lunar crescent visibility table, as both share the same parameter of elongation and rate of success. Al-Khazini table of lunar crescent visibility suggests that at a certain angular speed of the moon V_m, lunar crescent is visible at elongation more than the value of e_1 and visible before sunset if the arc of light is more than the value of e_2. A larger value of V_m indicates a longer earth–moon distance, thus making the threshold values of e_1 and e_2 to be bigger.

2.4.2.7 Nasirudin Al-Tusi Lunar Crescent Visibility Criterion

Muhammad ibn Muhammad ibn al-Hasan al-Tūsī was born in Tus, Iran, in 1201. His was famously known as Nasirudin Al-Tusi and an expert in various subjects, including astronomy, biology, and chemistry. In 1242, he compiles a collection of astronomy tables in handbook called *Ilkhani Zij*. *Ilkhani Zij* was dedicated to his patron, Ilkhan Hulagu. The handbook is a result of research and collaboration with astronomers at the Maragha observatory.[31]

Ilkhani Zij lunar crescent visibility uses two parameter: Sun–moon longitude and arc of separation or lag time. Similar to the Al-Khazini lunar crescent visibility table, Al-Tusi designed his table in the form of categorization. However, Al-Tusi provides a different presentation by introducing the condition of the visibility. The visibility categorization is expressed in terms of thin, moderate, bright, visible, and quite visible.

His lunar crescent visibility condition suggests that at sun–moon longitude less than 10 degrees, the lunar crescent will be visible at lag time more than 48 minutes. For lunar crescent that has sun–moon longitude more than 10 degrees, the crescent is confident to be visible at lag time more than 72 minutes, while the visibility of the lunar crescent will be difficult at lag time less than 52 minutes. Al-Tusi work on lunar crescent visibility is among the pioneer that categorized the lunar crescent visibility based on the thickness and visibility, a categorization that was then followed by Bruin, 500 years later.

2.4.2.8 Ibn Ayyub Al-Tabari Lunar Crescent Visibility Criterion

Abu Jaafar Muhammad Ibn Ayyub al-Tabari is a Persian astronomer. There is not much known about this life outside from his astronomical tables and works, which primarily written in Persian language. Ibn Ayyub al-Tabari's work on astronomical tables is noted to be one of the oldest works on astronomy written in Persian language by S. E. Kennedy. Ibn Ayyub al-Tabari published an astronomical handbook called *Mufrad Zij*. His astronomical handbook does not accompanied by explanatory text on how the table is calculated, and its mathematical origin. Therefore, modern historian unable to thoroughly examine the mathematical structure of *Mufrad Zij*.[32]

Mufrad Zij portrays a relationship between solar longitude and arc of separation. The values are calculated at the obliquity of 23°51′, declination of 9°30′, with location reference of 35°50′ of latitude. Therefore, it can be inferred that *Mufrad Zij* is produced with Rhages, or currently known as Ray, Iran, or Tehran as a location reference. *Mufrad Zij* lunar crescent visibility tables have interval of 10 degrees of solar longitude, with arc of separation maximum at 19°05′ or 76.33 minutes at 160 degrees of solar longitude, and minimum arc of separation at 38.4 minutes at 35 degrees of solar longitude. *Mufrad Zij* lunar crescent visibility table demonstrates influences from al-Khwarizmi and Indian planetary theories. *Mufrad Zij*, however, is more extensive by laying out the arc of separation for each 10 degrees of solar depression.

2.4.2.9 Al-Sanjufini Lunar Crescent Visibility Criterion

In 1366, Abu Muhammad al-Sanjufini computed a handbook as a gift to his patron, Prince Randa, a Mongol viceroy of Tibet and a direct descendant in the seventh generation from Genghis Khan. One of the 42 tables of the handbook is specified for predicting lunar crescent visibility. The table is computed for latitude 38.10 degrees, indicating that the table was computed for the second Mongol capital Yung-ch'ang fu. Al-Sanjufini table gives visibility function for each degree of lunar latitude for every 10 degrees of solar longitude.

His table suggests that a lunar crescent cannot be sighted before 14.5 hours of moon age. Al-Sanjufini lunar crescent visibility table does not yield satisfactory results in predicting lunar crescent visibility criterion, with 8.1 per cent and 35.7 per cent of the positive and negative observations contradiction rate in predicting visibility. It is interesting to note that the introduction of the lunar latitude as an additional parameter, in comparison to al-Khawarizmi lunar crescent visibility table, does not improve the reliability of al-Sanjufini visibility table.

2.4.2.10 Al-Lathiqi Lunar Crescent Visibility Criterion

Muhamad al-Lathiqi is a Muslim astronomy born around seventeenth century. Al-Lathiqi is a Syrian astronomer, with Lattakia to be his hometown. In 1698, al-Lathiqi published a table of lunar crescent visibility criterion, which use solar longitude and arc of separation as parameters, a similar design with al-Khawarizmi lunar crescent visibility table.

Al-Lathiqi lunar crescent visibility table methodology is to compute the difference in longitude between two luminaries at two-thirds of an hour after sunset. Then if the difference in longitude is greater than or equal to the visibility function limit, the lunar crescent is predicted to be visible. However, if the difference in longitude is less than visibility function limit, lunar crescent is predicted to be invisible.

Since it uses the same al-Khawarizmi design, this criterion has high percentages, making it an unreliable criterion. Al-Lathiqi criterion wrongly predicts 37 negative observations and 20 positive observations. Al-Lathiqi lunar crescent visibility table is slightly better than al-Khawarizmi lunar crescent visibility in predicting visibility; however, it is still under a range of unreliable lunar crescent visibility criteria.[33]

2.4.3 Modern Lunar Crescent Visibility Criterion

Study on the visibility of lunar crescent diminish after sixteenth century, parallel with the fall of Islamic science during the Middle Age.[34] Since that time, the new Hijri month is determined either with lunar crescent sighting or simple 29th–30th alternate rule. Research for lunar crescent visibility limit unable to gather interest until twentieth century.[35] In 1910, Fotheringham ignited the interest for lunar crescent visibility limit research, and it was followed by Maunder in 1911 and Danjon in 1936. The interest then rekindled among Muslim community, sparked by the conflicting lunar crescent visibility report and different date for the new Hijri month.[36] This led to the first Muslim lunar crescent visibility criterion since the Middle Age era, Istanbul Declaration in 1976,[37] which then followed by Ilyas series of lunar crescent visibility criterion, MABIMS lunar crescent visibility criterion in 1991 and Fatoohi lunar crescent visibility criterion in 1998.

The modern lunar crescent visibility criterion demonstrates a frequent use of altitude, azimuth and elongation parameters. Modern lunar crescent visibility criteria also saw the introduction of width and contrast thresholds to further increase the accuracy of lunar crescent visibility predictions. Modern lunar crescent visibility criteria are more composite in their parameters, and their design constructed through larger compilation of lunar crescent visibility reports, in contrast to Middle Age lunar crescent visibility criterion.

2.4.3.1 Fotheringham's Lunar Crescent Visibility Criterion

John Knight Fotheringham was born in 1874 in Britain. He was an expert in historical astronomy. He was also influential in establishing the Babylonian empire chronology. In 1910, Fotheringham published a study on the lunar crescent visibility criterion, incidentally, sparking interest in the matter, which has been stagnant for at least two centuries. While works of lunar crescent visibility have been published since 1868 by Johann Schmidt, the majority of the research discusses about the report of lunar crescent visibility and lunar crescent visibility criterion of the past, whereas Fotheringham was the first to introduce his own lunar crescent visibility.

Fotheringham incorporates altitude and azimuth for his lunar crescent visibility criterion.[38] His curve is calculated at sunset. Fotheringham does not state where he gathers ideas to construct the lunar crescent visibility curve using altitude and azimuth parameters, although Ilyas claimed that it was inspired by Battani lunar crescent visibility curve.[39] Fotheringham formulated his lunar crescent visibility curve from 55 positive data points and 21 negative data points of lunar crescent observation. His data is compiled from the collection of Mommsen and Julius Schmidt lunar crescent visibility data.[40] He further adds that his lunar crescent visibility curve is applicable at any given location, with slight adjustment according to atmospheric extinction.

His curve can roughly be represented in the expression of

$$Arcv = -0.1758223322\nu AZ + 0.0225942071\Delta AZ^2$$
$$+ -0.0009955850\Delta AZ^3 + 12.0825223783 \qquad 2.1$$

Source: Researcher Data.

Fotheringham lunar crescent visibility criterion suggests that the lunar crescent will be visible at lower altitude if it separated by considerable number of azimuths. Deducing from his visibility curve expression, at 38 degrees of azimuth, lunar crescent is visible at 0 degrees of altitude. In real observation, lunar crescent is unfeasible to be visible at 0 degrees of altitude due to the effect of concentrated air mass and high-level atmospheric extinction.[41]

Maunder criticized the Fatoohi visibility curve, stating his design is primarily based on positive lunar crescent visibility records and ignored the majority of negative sighting. Maunder adds that a lunar crescent is reported to be visible at a given parameter but does not guarantee it is visible at other time and location. This is because clouds and atmospheric conditions could hamper visibility of a lunar crescent.[42] Fotheringham's inattention to negative lunar crescent observation makes its visibility curve to be located at a higher visibility threshold, consequently unable to accurately predict several negative lunar crescent visibility reports.[43]

Fotheringham data compilation also only collected from two source: Mommsen and Schmidt, which located at Athens. This making his visibility only viable for Athens and susceptible to error at other latitudes. His data of altitude and azimuth is calculated without consideration of parallax and refraction, making his data subject to error up to 1 degree in real observation. Fatoohi, when examining Fotheringham lunar crescent visibility criterion, found out that Maunder claimed Fotheringham lunar crescent visibility criterion flaw is true. Fotheringham high visibility curve ignores most of the negative lunar crescent sighting. In addition, Fotheringham claims that his criterion is adaptable at any given latitude is erroneous, as it is found out that it has a high error in predicting lunar crescent visibility at other latitudes.[44]

Fatoohi lunar crescent visibility criterion, despite its deficiency, has sparked a positive competition among astronomers in designing lunar crescent visibility criterion. His altitude and azimuth parameters led to the design of Maunder lunar crescent visibility criterion, and inspired research on other topocentric parameters of lunar crescent visibility, such as elongation and width[45] His framework of altitude–azimuth parameters is still being used today by Muslim countries such as MABIMS to determine their new Hijri month.

2.4.3.2 Maunder's Lunar Crescent Visibility Criterion

Edward Walter Maunder was a British resident, born in 1851. He was an influential astronomer in solar observation, famously associated with the term *Maunder Minimum* to describe the period of prolonged solar minimum from 1645 to 1715.[46] In 1911, Maunder published a lunar crescent visibility criterion in his article 'On the Smallest Visible Phase of the Moon.' His article is a form of refinement from the works of Fotheringham, as he heavily criticizes Fotheringham's works as being too pessimistic in predicting lunar crescent visibility. His lunar crescent visibility criterion is demonstrated with a table and can be expressed in the form of the following formula:

$$ArcV = 11 - 0.05\Delta AZ - 0.01\Delta AZ^2 \qquad\qquad 2.2$$

Source: Researcher Data.

Similarly, with Fotheringham, Maunder incorporates altitude and azimuth for his lunar crescent visibility criterion. Maunder criterion is calculated at sunset. Maunder applies the same Fotheringham's altitude–azimuth framework for his lunar crescent visibility criterion, with the only difference is that Maunder visibility curve is lower than Fotheringham visibility curve. This is due to the consideration of negative observation in Maunder visibility curve.[47]

Maunder lunar crescent visibility criterion uses the same data from Fotheringham data with additional 11 more data from various latitudes, amassing 87 data points of lunar crescent visibility records, 66 positive and 21 negative observations. Maunder lunar crescent visibility data is clustered around the location of Athens. Maunder insists that his visibility curve is more reliable since it considers both negative and positive observation. As Maunder lunar crescent visibility curve uses the same framework with Fotheringham, his criterion would suggest that a lunar crescent is visible at 0 degrees of arc of vision and 30 degrees of difference in azimuth, which would be impossible due to atmospheric extinction and air mass. Alternatively, Maunder visibility curve suggests that the lunar crescent is visible at a 11.0 degree arc of vision when its azimuthal difference is 0 degree. This is not necessarily the case, as Fatoohi recorded a visible lunar crescent at as low as 6.2 degrees and 0.5 azimuthal different.[48]

Maunder correction on Fotheringham visibility curve, although applaudable, is still under the framework of altitude–azimuth visibility curve. Despite Maunder's attempt to accurately draw the line between positive and negative lunar crescents his altitude–azimuth criterion has produced the same issue with Fotheringham criterion. A high visibility curve would favour prediction for negative data but unable to predict positive data, while an attempt to lower the visibility curve will favour prediction for positive data however reduce the successful rate of negative data prediction. It is also noted that the framework of altitude–azimuth lunar crescent visibility criterion is latitude dependent.

Fatoohi, in his assessment of Maunder visibility curve accuracy, found out that it has contradiction rate of 17.8 per cent in predicting positive observation, while 15.5 per cent of the negative lunar crescent observation data fall above Maunder visibility. Maunder visibility curve is far superior to other visibility curves that adopt the altitude–azimuth framework, such as Neugebauer and Schoch; however, due to its adoption of the altitude–azimuth framework itself, it capability to predict lunar crescent visibility is not satisfactory.[49]

Maunder lunar crescent visibility criterion, despite its flaw, demonstrated how lunar crescent visibility is designed. Maunder shows that rather than favouring positive observation in the construction of lunar crescent visibility criterion, the consideration of negative observation greatly increases the accuracy of any criterion.

2.4.3.3 Danjon's Lunar Crescent Visibility Criterion

André-Louis Danjon was an astronomer born in Caen, France, in 1890. He was a notable French astronomer, famously credited for the introducing qualitative scale for measuring the appearance and luminosity of the lunar crescent, currently known as the Danjon scale.[50] In 1936, through a collection of 75 measurements and estimation of lunar crescent lengths, Danjon publishes a work entitled 'Le Croissant Lunaire,' which is translated as 'The Lunar Crescent'. In his work, Danjon explains a relationship between the angle of separation from the sun and moon or elongation against the length of a lunar crescent. Danjon states that the length of lunar crescent increases from 0 degrees to 180 degrees in proportional with elongation from 7 degrees to 180 degrees. He deduced that the lunar crescent is invisible for elongation below 7 degrees due to being shadowed by the lunar mountain.

Danjon compilation of 75 measurements of lunar crescent length has elongation ranging from 8 degrees to 90 degrees. This indicates that the value of 7 degrees is not the result of direct measurement of lunar crescent; instead, it is a product of extrapolation from his graph. In fact, the lowest crescent length is recorded at 6.2 degrees with 8 degrees of elongation. This means that the 7-degree result is an interpretation by Danjon and subjected to other interpretations by other researchers.

The limit of 7-degree elongation for crescent length, or currently known as the Danjon limit, is highly contested by researchers. McNally has argued that the average lunar radius has a variation of 0.6 per cent, a variation that is too small to cast a shadow that overcasts lunar crescent length.[51] He then explains that the deficiency of lunar crescent length is due to atmospheric seeing on cusp of the lunar crescent. Atmospheric seeing causes the cusp brightness of the lunar crescent to be reduced, hence impacting on shorten the visible cusps of lunar crescent.

Schaefer, on the other hand, provided a different explanation on the shortening of the lunar crescent.[52] First, he agreed with McNally that it is not plausible to attribute lunar crescent shortening with shadow of lunar mountain, as it requires a height of 12 km of lunar mountain shadow to overcast the lunar crescent. The highest mountain on the moon is Mons Huygens that has 5.5 km in elevation.[53] However, Schaefer disagrees with McNally on the causative effect of atmospheric seeing on the length of lunar crescent. Schaefer supplements his disagreement by stating that his Moonwatch project indicates that both telescopic and visual observation report a same length of lunar crescent. By McNally modelling, telescopic and visual observers should have different impacts of atmospheric seeing, thus contributing to different lengths of lunar crescent.[54] Schaefer concedes that McNally modelling is not applicable for explaining the length of lunar crescent. Schaefer then suggested that the reason of the shortening length of lunar crescent at lower elongation is due to the sharp reduction of integrated brightness towards the cusps. The reduction in brightness decreases the detectable contrast, thus contributing to the shortening of lunar crescent. Schaefer then cements that, with his theory, a 7.5 degree of elongation would be a plausible elongation limit for detectable lunar crescent or Danjon limit.

Ilyas agreeing with Schaefer, concedes that the shortening is due to the brightness deficiency at the cusps, making it undetectable in human eye.[55] However, Ilyas provided a different model to explain his theory. Ilyas eliminated the 8-degree elongation data in Danjon measurement, and then provided a new extrapolation curve that has the lowest limit of 10.5 degrees of elongation. He also supported his argument for 10.5 degrees in elongation by deriving the value of elongation from the lowest limit of width. He argues that the lowest limit of detectable width $w = 0.25'$ would attribute to elongation of 10.5 degrees.

Sultan attempted to provide a different explanation for the shortening of lunar crescent length.[56] He argues lunar crescent visibility is dependent on the surface brightness per area of the lunar crescent instead of total integrated brightness. This means that the absence of lunar crescent data for elongation below 7.5 is due to the surface brightness of the lunar crescent at the cusps having a low contrast to be visible with naked eye.[57] For optical-aided observation, however, are in different cases. Sultan argues that optical-aided observation is able to break the 7-egree limit of Danjon as optical-aided observation able to increase the size of the lunar disc, while maintaining its surface brightness. Sultan then proved theoretically that optical-aided observation at 200 magnifications was able to observe a lunar crescent at 5 degree elongation.

Hasanzadeh conducted a multi-test to examine the Danjon limit of lunar crescent visibility.[58] Amir experiments involved extrapolation of elongation against length of lunar crescent, with additional parameters of atmospheric seeing, lunar mountain shadowing and libration. Amir also experimented with the Sultan method of determining Danjon limit by observing the lunar crescent at 120 magnifications. Interestingly, all Hasanzadeh experiments result in a limiting elongation of 5 degrees for lunar crescent visibility.

Despite the differences in explaining the reason for lunar crescent length shortening, Danjon, McNally, Schaefer, Ilyas, Sultan, and Hasanzadeh contributed to understanding the limits of lunar crescent visibility. Danjon and Schaefer are correct to predict the visibility limit at 7–7.5 degree elongation, as it is proven that the naked eye is capable of detecting lunar crescent at 7.7 degree of elongation. McNally, Sultan, and Amir argue that lunar crescent is possible to be sighted at elongation below 7 degrees are warranted, with the 6.0 degrees and 3.5 degrees of elongation for optical-aided and CCD observation. Ilyas claim for 10.5 degree elongation for naked eye limit is somewhat justified, with the majority of the lunar crescent visibility fall under the range of 9.0–10.5 degree of elongation.

2.4.3.4 Bruin's Lunar Crescent Visibility Criterion

Frans Bruin was born in 1922 in Hague, Netherlands. He was a professor of physics at American University of Beirut and director of the Universität Bern Astronomical Institute observatory. He was one of the famous historians of astronomy and has luxury of working together with the likes of Otto Neugebauer and Edward Kennedy.[59] In 1977, Bruin constructed a lunar crescent visibility criterion that pioneered the inclusion of astrophysical aspect of lunar crescent visibility.

Bruin incorporates the parameters of lunar width, altitude, and azimuth in his criterion. His criterion is expressed in the various values of width ranging from 0.5′, 0.7′, 1′, 2′, and 3′, with attribution of solar depression and arc of vision on its axis. Bruin lunar crescent visibility criterion is expressed in Equation 2.3.

$$ArcV = -0.1324039674w + 0.0009057913w^2$$
$$+ -0.0000021108w^3 + 11.5621745317 \qquad 2.3$$

Source: Researcher Data.

The application of Bruin lunar crescent visibility criterion is complicated. First, the width of the observed lunar crescent needs to be calculated first. Taking an example lunar crescent width of 2′, during lunar crescent observation, at 5.5 degrees of lunar altitude, lunar crescent is visible at solar depression of 4.0 degrees until 0.8 degrees, meaning that it has 12.8 minute windows of opportunity. Bruin lunar crescent visibility criterion is not only able to predict the visibility of the lunar crescent, it is also at the same time able to estimate the time windows for successful observation.

Bruin, in designing his criterion, has the following assumption: First, he assumes that the sky brightness is uniform regardless of altitude and azimuth, with only solar depression as a single brightness variable.[60] Second, Bruin assumes that the brightness of the lunar crescent is uniform across its surface, with only lunar crescent altitude acts as a presenter for atmospheric extinction.[61] Third, Bruin assumes that minimum require contrast for lunar crescent visibility are associated with lunar surface area. For this assumption, Bruin adopted the work of Siedentopf circular disc visibility threshold and convert it into lunar width.[62] Bruin uses assumption in his design for lunar crescent visibility criterion, so he does not use based on actual observation of lunar crescent. Bruin stated that his criterion has been experimented for 10 years, and his assumptions are correct without requiring further refinement.

All three Bruin assumptions are incorrect. First, the assumption that sky brightness is uniform will single solar depression act as a brightness variable is entirely wrong. Kastner has developed a modelling that demonstrates the brightness of the sky during twilight is dependent on solar depression, altitude, and azimuth of the observed object.[63] Kastner modelling warrants a high accuracy and is still relevant for the current application, which adopted by Faid et al. for his modelling of light-polluted twilight sky brightness.[64] Second, the assumption that brightness of the lunar crescent to be singularity dependent to lunar crescent altitude is not entirely correct. Although lunar crescent altitude can represent atmospheric extinction in the simplest form, the impact of atmospheric extinction on lunar brightness is more complex and requires complex variables. Schaefer has laid out the computation required to measure the impact of atmospheric extinction on lunar brightness, encompassing air mass, temperature, season, atmospheric layer, humidity, altitude, latitude, and wavelength. Thus, to simply express the impact of atmospheric extinction on lunar brightness in the form of lunar crescent altitude is an oversimplification. Third, Bruin adopted Siedentopf circular disc visibility threshold in his criterion by assuming it was applicable for lunar crescent visibility threshold. Circular disc visibility and lunar crescent visibility threshold are heterogenous. This is because the surface area and the shape of lunar crescent are entirely different with circular disc. Blackwell's model of crescent visibility threshold in 1946 is more suitable for Bruin lunar crescent visibility criterion instead of Siedentopf works.[65]

Fatoohi, in assessing the reliability of Bruin lunar crescent visibility criterion, has found out that Bruin has underestimated the capability of human eye to detect the limiting width of lunar crescent.[66] There are 77 reports of lunar sighting with lunar crescent width less than 0.5′ observed by naked eye, with thinnest to be at 0.17′. This is way below the visibility limit of Bruin lunar crescent width of 0.5′. Fatoohi further adds that Bruin that miss predicts 27.7 per cent of the positive observation, and 9.6 per cent of negative observation.

Bruin lunar crescent visibility criterion, despite the incorrect assumption of lunar crescent visibility and its underestimation of human eye detection capability, was a pioneer in designing an astrophysical lunar crescent visibility criterion. Bruin criterion creates a pathway for Schaefer, Sultan, and Faid to create their own astrophysical lunar crescent visibility criteria.[67]

2.4.3.5 Ilyas's Lunar Crescent Visibility Criterion

Ilyas was born in Meerut, India, in 1950. He is one of the Muslim pioneers in research on lunar crescent visibility. In 1983–1994, he has published at least ten articles on lunar crescent visibility and Islamic calendar. Ilyas was at the forefront of bringing Muslim astronomers into research of lunar crescent visibility and the Islamic calendar, where during his time the majority of these endeavours were carried out by Islamic scholar without prior scientific or astronomical knowledge.[68] His work on lunar crescent visibility and the Islamic calendar sparks the interest of other Muslim astronomers to research and examine the reliability of lunar crescent visibility criterion.

Ilyas has produced various lunar crescent visibility criteria, moon age-latitude lunar crescent visibility criteria, lag time-latitude lunar crescent visibility criterion, lunar crescent altitude and sun–moon azimuth, revision of Danjon limit, revision of Bruin lunar width, arc of light, and arc of vision. Ilyas primary lunar crescent visibility criterion is altitude–azimuth criterion.

$$Arcv = -0.0027356815\,DAZ + -0.0136648716\,DAZ^2$$
$$+0.0002119205\,DAZ^3 + 10.2832719598 \qquad\qquad 2.4$$

Source: Researcher Data.

Ilyas suggested that the limiting elongation for lunar crescent visibility is 10.5 degrees, 3.5 degrees more than Danjon limit, while the limiting altitude for lunar crescent visibility is 10 degrees. Ilyas derived the lunar crescent altitude values from Maunder lunar crescent visibility criterion, where Maunder limiting threshold of lunar crescent altitude at 0 degrees of azimuth is 11.0 degrees.[69] The 10.5 elongation limit is derived from a re-extrapolation of Danjon graph and a reinterpretation of Bruin limit of lunar width.[70] Ilyas found out that if the Bruin lunar width limit is lowered to 0.25′, it would correspond to the geocentric elongation of 10.5 degrees. Ilyas claimed that the drawing of his lunar crescent visibility graph is a combination of Bruin and Maunder lunar crescent visibility criterion. Ilyas further cemented that his criterion is agreeable at any given latitude and contains a small value of uncertainty.

Fatoohi argues that Ilyas has suffered a fundamental flaw in his design of lunar crescent visibility criterion.[71] Maunder and Bruin lunar crescent visibility criterion are not related to one another. Maunder derived his lunar crescent visibility criterion from 91 data points of lunar crescent visibility, while Bruin designed his lunar crescent visibility criterion from the theoretical values of sky brightness, lunar crescent illumination, and contrast threshold. Ilyas also did not make any attempt to demonstrate how the combination of two independent lunar crescent visibility criterion works. In addition, it is stipulated by Fatoohi that Maunder lunar crescent visibility criterion does not work at all latitudes, and Bruin lunar crescent visibility criterion has extensive range of uncertainty. In addition, McPartlan commented that Ilyas elongation limits to underestimation of human eye and should be lowered by 0.5 degree to account for the number of positive lunar crescent observations that fall under Ilyas invisibility line.[72]

Fatoohi has found out that the Ilyas altitude-elongation criterion has a 29.8 per cent contradiction rate in predicting invisibility, and a 7.8 per cent contradiction rate in predicting visibility. Ilyas altitude–azimuth lunar crescent visibility criterion unable to predict 28.6 per cent of negative sighting and 11.3 per cent positive sighting. His limiting value of elongation is also not dependable, as it is found out that the lunar crescent is detectable at 7.7 degrees in elongation. Despite Ilyas flaws in his design of lunar crescent visibility criterion, Ilyas has shown to the possibility various presentations in designing lunar crescent visibility criterion. Ilyas also perhaps the most influential astronomer in lunar crescent visibility research among Muslims.

2.4.3.6 Schaefer's Lunar Crescent Visibility Criterion

Bradley Schaefer is a Professor Emeritus at Louisiana State University. He was awarded with the Nobel Prize and Gruber Prize for his team's discovery of dark energy. In 1983, he embarked on a journey of lunar crescent visibility research. In his publication on lunar crescent visibility entitled, 'Algorithm of Lunar Crescent Visibility,' he criticized the current lunar crescent visibility criterion to be limited to geometrical measurement, whereas visibility is a problem that involves atmospheric and human eye sensitivity.[73] He proposed that the lunar crescent visibility criterion should be designed with consideration of physical, meteorological and physiological equations, in the same framework that Bruin has designed his lunar crescent visibility criterion.[74]

To achieve his goal of comprehensive lunar crescent visibility criterion, Bradley Schaefer launched the Moonwatch project, an open project for lunar crescent observation. Through the project, Schaefer was able to gather data on the visibility of lunar crescent through instrumentation, physiological and atmospheric perspectives.[75] He also conducted studies on lunar brightness,[76] twilight sky brightness,[77] atmospheric extinction,[78] telescopic limiting magnitude,[79] lunar physical observation, and visibility threshold to further refine his algorithm.[80] Schaefer tested his algorithm on sunspot visibility and validating the date of Jesus crucifixion.[81] In 1993, Schaefer produced his computation formula for his algorithm,[82] and in 2000, he updated the final version of his algorithm computation formula.[83]

Numerous researchers have tried to emulate Schaefer lunar crescent visibility algorithm. Fatoohi commented that Schaefer never actually published the full version of his formulation, despite comparatively assessing his algorithm to other lunar crescent visibility criterion.[84] Ilyas highlighted that Schaefer's lunar crescent visibility algorithm was not published to the public during his time. He also added that the Schaefer algorithm is too complicated and not practical for long-term prediction, particularly in the application of Islamic calendar determination.[85] Yallop, also in agreement with Fatoohi and Ilyas, noted that replicating Schaefer-calculated algorithm result is difficult as the information regarding to the algorithm is conflicting between his papers.[86]. In partial assessment of Schaefer algorithm, Loewinger has found out that Schaefer miscalculated the lag time value of his data.[87] Sultan also found that Schaefer has a confusing definition of lunar brightness, interchanging the definition of integrated brightness and surface brightness frequently.[88] Faid, found Schaefer has miscalculated some of value; however, after correction, it was able to perform at an accuracy of 73 per cent in predicting negative sighting visibility of lunar crescent, and 93 per cent for positive sighting.[89] Schaefer algorithm, despite its flaws indicate that the approach to convert the computation of lunar crescent visibility into a full theoretical formulation is entirely plausible.

2.4.3.7 Yallop's Lunar Crescent Visibility Criterion

Bernard Yallop is a British-born astronomer who is positioned as a Superintendent of the Nautical Almanac, an almanac that is responsible for producing astronomical data for astronomers, mariners, aviators, and surveyors. Bernard Yallop contributed in to determining the Muslim prayer time of Isha and Subh by defining the beginning of twilight at 18 degrees solar depression,[90] a value that has been used by the majority of Muslim countries at the moment[91]. In 1988, Yallop published a lunar crescent visibility criteria that can be categorized into ranges of visibility.

Yallop formulated his lunar crescent visibility criterion by adapting the arc of vision and sun–moon azimuth into q value expression. The formulation is expressed in Equation 2.5.

$$q = \frac{arcV - (11.8371 - 6.3226w + 0.7319w^2 - 0.1019w^3)}{10} \qquad 2.5$$

Source: Researcher Data.

Yallop categorized the values of q into ranges of lunar crescent visibility, which are easily visible, visible under perfect condition, may need optical aid to find crescent, will need optical aid to find crescent, not visible with a telescope and not visible. Yallop based his Q formulation from Indian lunar crescent visibility criterion, and Neugebauer lunar crescent visibility criterion.[92] Yallop categorization of visibility ranges is based on 295 records of lunar crescent observation compiled by Schaefer and Doggett.[93]

Fatoohi criticized that Yallop categorization, particularly E and F categories, is not needed as both can be counted as invisible sightings. Fatoohi added that Neugebauer and Bruin lunar crescent visibility criterion is not dependable enough to function as a basis for Yallop lunar crescent visibility criterion. Fatoohi also argued that Yallop categorization is not balanced, as he found out that from 295 points data of lunar crescent sighting, 166 fall under category A, while only 68, 26, 14, 4, and 17 fall under categories B, C, D, E, and F, respectively. These disparities indicate more data is needed for each category to validate the ranges of visibility.[94]

In examining the Yallop criterion, Fatoohi finds a high number of errors, except for category A. Fatoohi notes that Yallop criterion is highly unreliable as it is unable to accurately predict the lunar crescent visibility.

Yallop lunar crescent visibility criterion is heavily criticized by Fatoohi for its adoption of inconsistence visibility ranges, and weak mathematical foundation. Despite its weakness, Yallop is attributed for popularizing the concept of visibility ranges and best time, which then influenced researchers like Qureshi and Sultan to produce their definitions of visibility ranges and best time, respectively.[95]

2.4.3.8 Fatoohi's Lunar Crescent Visibility Criterion

Louay Fatoohi is a Baghdad-born Muslim who currently resides in Birmingham England. He is husband to Dr Seetha al-Dargazelli, a Professor at Durham University and Aston University. He has a degree in physics from Baghdad University, Iraq, and a Doctor of Philosophy in Astronomy from Durham University in 1998. His PhD thesis is entitled 'First visibility of the lunar crescent and other problems in historical astronomy'.

In his thesis, lunar crescent altitude that is located in between $ArcV_{UpperLimit}$ and $ArcV_{LowerLimit}$ is located at a zone of uncertainty. Fatoohi developed the idea of zone of uncertainty from Ilyas into his design of lunar crescent visibility criterion. While Ilyas idea of zone of uncertainty refers to the geographical latitude–longitude, where the lunar crescent cannot accurately predict using his criterion in the International Lunar Date Line,[96] Fatoohi incorporate the zone of uncertainty directly in his formulation of lunar crescent visibility criterion, where the visibility of the lunar crescent cannot be predicted accurately.

Fatoohi noted that both upper limit and lower limit of his criterion only have an error of 5.9 per cent for predicting negative lunar crescent sighting, and 3.6 per cent for predicting positive lunar crescent sighting; however, in between both limits, the accuracy of lunar crescent prediction fall rapidly. In his test, he found out that the zone of uncertainty in his criterion accommodates 27.4 per cent of negative sighting error and 16.4 per cent of positive sighting error.

He notes that the implementation of his lunar crescent visibility criterion is more applicable for determination of the new Hijri month, where both the practice of astronomical calculation and lunar crescent sighting can be adopted. If the lunar crescent is located above the upper limit of the Fatoohi lunar crescent visibility criterion, then confidently lunar crescent will be sighted, and the new Hijri month has commenced. If the lunar crescent is located below the lower limit of the Fatoohi lunar crescent visibility criterion, then, confidently the lunar crescent will not be sighted, and the current Islamic month will be continued until 30th day. However, because the lunar crescent is located in between the upper and lower limits, it needs to be sighted to confirm its visibility. Fatoohi notes that in his test on 300 months of Islamic calendar in Mecca, Baghdad and Casablanca, there are around 7 per cent of cases where the lunar crescent is located in the zone of uncertainty.

The idea of using lunar crescent sighting for zone of uncertainty on paper looks practical and relatable to the practice of the Prophet. However, in real cases, as discussed in the previous chapter, some countries have limitations in exercising lunar crescent sighting to determine the new Hijri month, making the idea of incorporating the zone of uncertainty on lunar crescent visibility criterion not feasibile for real-time application. The incorporation of zone of uncertainty also limits the usage of the lunar crescent visibility criterion for historic calendrical dating purposes, as lunar crescent that are located at the zone of uncertainty cannot accurately calculated.

2.4.3.9 Odeh's Lunar Crescent Visibility Criterion

Mohamad Shaukat Odeh was born in Kuwait in 1979. He is a member of the Arab Union for Astronomy and Space Science. He attributed for founding the global lunar crescent observation project, called the Islamic Crescent Observation Project, since 1998. The project since then has collected more than 2,000 data points of lunar crescent sighting worldwide. Odeh is also involved in the development of Accurate Time, an Islamic astronomy software that can function as a lunar crescent visibility calculator, an Islamic calendar calculator, and a prayer time calculator. He is one of the most influential lunar crescent sighting astronomers of our time.

Odeh in 2005 published a lunar crescent visibility criterion. His lunar crescent visibility criterion is categorized into ranges of visibility, similar to the Yallop lunar crescent visibility criterion. Odeh determines his V value using the same formula as how Yallop determines his Q value. Odeh lowered his arc of vision threshold from 11.8371 degrees into 7.1651 degrees.

Odeh lunar crescent visibility criterion is formulated through the compilation of 737 records of lunar crescent sighting. These include 294 records of lunar crescent sighting from Schaefer list, 6 records from Jim Stamm, 42 records from the South Africa Astronomical Observatory, 15 records from Mohsen Mirsaeed, 57 records from Alireza Mehrani, and 323 records from ICOP. As Odeh follows Yallop lunar crescent visibility criterion, it has the same weakness as Yallop criterion. Although Odeh visibility classification is more practical than Yallop visibility classification, the number of lunar crescent sightings on each visibility group is not balanced with 46 lunar crescent sightings on Group D, 117 lunar crescent sightings on Group C, 255 lunar crescent sightings Group B, and 160 lunar crescent sightings on Group A. Furthermore, the ratio of positive to negative sightings on each group is lopsided, with Group A having majority of negative lunar crescent sightings, and group D having majority of positive sightings.

2.4.3.10 *Qureshi's Lunar Crescent Visibility Criterion*

Mohamad Shahid Qureshi is a Pakistan born astronomer, mathematician, and astrophysicist. He is former director and Professor at the Institute of Space and Planetary Astrophysics, Karachi Universiti Pakistan. Shahid Qureshi's Doctoral thesis is entitled 'Earliest Visibility of Lunar Crescent' making him an expert in designing lunar crescent visibility.[97] Shahid Qureshi has published at least five papers concerning the visibility of lunar crescent in Pakistan.

Qureshi produced his own lunar crescent visibility criterion in 2010. Shahid Qureshi criteria are in a similar framework with Yallop and Odeh criteria, which, by adapting ranges of lunar crescent visibility. Qureshi ranges of visibility are categorized into easily visible, visible under perfect condition, may require optical aid to find crescent, require optical aid and not visible with optical aid. Qureshi visibility ranges are based on s value. The s value takes the same arc of vision and width parameter used by Yallop and Odeh; however, Qureshi changed the coefficient to fit with his data. Qureshi data is calculated using some website that claimed to adapt Schaefer visibility logarithm. The website is however inaccessible to verify the computation. Qureshi highlights that his s value is more accurate as it considers the brightness of the sky, lunar crescent illumination, and detectable contrast threshold; however, he does not demonstrate how the formulation of the s value is conducted. As Qureshi shares the same lunar crescent visibility criterion style as Yallop and Odeh, it does have the issue of unbalanced lunar crescent sighting data of each visibility group.

2.4.3.11 Caldwell's Lunar Crescent Visibility Criterion

John Caldwell is an astronomer who graduated from California Institute of Technology in 1974. In 1979, he obtained his PhD from Princeton University. Previously he was a research fellow at the South African Astronomical Observatory in Cape Town, South Africa. In 2012, John Caldwell published a study on lunar crescent visibility criterion, on the Monthly Notes of the Astronomical Society of South Africa Journal.[98] To date, John Caldwell lunar crescent visibility criterion is the only published work on lunar crescent visibility criterion from African countries.

Caldwell incorporates moonset–sunset lag time and arc of light in his lunar crescent visibility criterion. Caldwell argues that lag time is a better parameter to determine lunar crescent visibility, as it is applicable at various degrees of latitude. This is because lag time is correlated to the separation angle between sun and moon, in contrast to altitude, where it is highly correlated to the local horizon. Lag time is also best paired with elongation, as both do not have a linear relationship to each other.

The Caldwell lunar crescent visibility criterion is based on 36 data points of positive naked eye and 58 data points of positive optical-aided lunar crescent sighting from various latitudes and longitudes. As the Caldwell lunar crescent visibility criterion is based on a small dataset, this makes his criterion susceptible to error. While in his graph Caldwell includes negative lunar crescent sighting, the details of negative lunar crescent sighting data are not included in this chapter, making reassessment of his criterion limited. Lag time parameter is also criticized by Ilyas, Schaefer, and Fatoohi to be highly unreliable and dependent to latitude. Caldwell criterion design also suffers from Yallop lunar crescent visibility design, which is an unbalanced number of lunar crescent sighting reports on each visibility categorization.

2.4.3.12 Krauss's Lunar Crescent Visibility Criterion

Rolf Krauss published a study in 2012 on lunar crescent visibility using Babylonian data of lunar crescent sighting. Krauss 95-page article contains arguments about data validity, interpretation of modern lunar crescent visibility criterion based on Babylonian data, effect of weather on lunar crescent visibility, and an azimuth–altitude lunar crescent visibility criteria.

Krauss includes the seasonal factor into his criterion, citing Schaefer visibility logarithm to support his inclusion; however, the inclusion causes Krauss's criterion to have a large deviation error, which is up to 1.8 degrees. For calendrical purposes, a large deviation error can lead to unreliable lunar crescent visibility criterion.

2.4.3.13 Gautschy's Lunar Crescent Visibility Criterion

Rita Gautschy is a lecturer, currently teaching in the Department of Environmental Science, University of Basel, Switzerland. She is an active member of the International Astronomical Union, currently affiliated with Division a Fundamental Astronomy IAU member. She completed her master's and doctoral degrees, in 1997 and 2001, respectively, under the topic of carbon stars. Since 2011, she has been delving on issue of archeoastronomy and history of astronomy, such as Egypt chronology and Medieval observation.[99] In 2014, Gautschy published a work on lunar crescent visibility criterion entitled 'On the Babylonian sighting-criterion for the lunar crescent and its implications for Egyptian lunar data'.[100] The article endeavours to produce a lunar crescent visibility criterion based on Babylonian prediction and lunar crescent visibility records and utilize the produced criterion in establishing an absolute Egyptian chronology.

Gautschy used Yallop lunar crescent visibility criterion to validate the Babylonian records of lunar crescent sighting. Records that contradict Yallop lunar crescent visibility criterion is recalculated to ensure their accuracy. Inaccurate or unclear records of lunar crescent sightings are rejected. Gautschy argues that Krauss judgement to design a lunar crescent visibility criterion based on season is not justified, as it is proven that season does not affect visibility of lunar crescent. Gautschy used parameters of lag time and difference in azimuth, as lag time is insensitive to difference calculation reference, either topocentric or geocentric. Gautschy also evaluated her criterion using the Odeh lunar crescent visibility criterion and found her result to follow the Odeh visibility prediction.

Gaustchy is able to provide a refresh outlook on lunar crescent visibility criterion based on Babylonian lunar crescent visibility records. While Kraus adopted modern altitude–azimuth lunar crescent visibility criterion, Gaustchy was adamant that lag time parameter, which has been adopted since the Babylonian age is just as efficient as other topocentric parameters. While she admitted that lag time is dependent on latitude, it is applicable for her research purpose, which is to produce an absolute Egyptian chronology specific for Egypt's latitude.

2.4.3.14 Alrefay's Lunar Crescent Visibility Criterion

Thamer Alrefay is an Assistant Professor from the Space Research Institute, King Abdul Aziz City for Science and Technology, Riyadh, Saudi Arabia. He is a member of the Canadian Association for Physicists, and completed his PhD at the University of New Brunswick, Canada, in 2014 under the subject of space physics. Thamer Alrefay interests are space physics, fireball observation and earth bow shock.

Alrefay and his fellow researchers of King Abdul Aziz City for Science and Technology published research in 2018 on the earliest visibility of the lunar crescent. The research is conducted based on 545 observations of lunar crescent sightings in Saudi Arabia within the duration of 27 years. Alrefay et al. developed a lunar crescent visibility criterion using width and arc of vision parameters, in similar fashion with Yallop, Qureshi and Odeh. Alrefay lunar crescent visibility criterion is classified into two ranges: Naked-eye and optical-aided observations.[101]

Alrefay argues that Yallop and Odeh lunar crescent visibility criteria are not consistent with other lunar crescent visibility data calculation references. In addition, Yallop and Odeh adopt a topocentric width parameter, without any argument why topocentric parameter would help in determining the limiting visibility of lunar crescent. Alrefay argues that his lunar crescent visibility criterion is based on geocentric lunar crescent parameter, and his criterion is consistent with other lunar crescent visibility data calculation references. Alrefay lunar crescent visibility criterion, however, is based on Saudi Arabia data and limited to 595 lunar crescent sightings, while Odeh lunar crescent visibility criterion is based on 737 worldwide lunar crescent sighting data. This makesAlrefay not dependable to determine the lunar crescent visibility outside Saudi Arabia. The contradiction rate analysis of Ahmad lunar crescent visibility criterion is not examined by any scholar so far.

2.4.4 Country-Based Lunar Crescent Visibility Criterion

A country-based lunar crescent visibility criterion is a criterion that is used for the purpose of a country Hijri calendrical determination. A country's based lunar crescent visibility criterion is usually designed at a lower line of a lunar crescent visibility, as it is to ensure that none of the lunar crescent is sighted below the criterion. In addition, a country-based lunar crescent visibility criterion is usually designed in a conditional style lunar crescent visibility criterion. This is contrast to research based lunar crescent visibility criteria such Alrefay et al. and Gautschy, where they used expression-style lunar crescent visibility criterion. It can be deduced that the reason for using conditional style lunar crescent visibility criterion is that it is simpler for Hijri calendrical calculation, while the equation style lunar crescent visibility criterion requires more computation power and advanced programming techniques to calculate Hijri calendar, particularly computations that involve long years of Hijri calendar.

2.4.4.1 Saudi's Lunar Crescent Visibility Criterion

Saudi Arabia is a country that houses Muslim pilgrimage locations, which are Mecca and Madinah. Muhamad was also born in Mecca, and Islam expansion has Mecca and Madinah as its epicentre. This makes Saudi Arabia the most influential country among Muslim community. Due to this reason, several countries closely follow the Saudi Arabian date of Hijri month. Dates of religious importance, such as day of Arafah and Eid Adha, impact Muslims worldwide as it relates to their religion practice. This makes a number of countries follow Saudi Arabia lunar crescent visibility criterion, such as Afghanistan, Albania, Algeria, Armenia, Austria, Azerbaijan, Bahrain, Belgium, Bolivia, Bulgaria, Burkina Faso, Chechnya, Denmark, Finland, Georgia, Hungary, Iceland, Iraq (Sunnis), Italy, Japan, Kazakhstan, Kuwait, Kyrgyzstan, Lebanon, Mauritania, Palestine, Philippines, Qatar, Romania, Russia, Sudan, Sweden, Switzerland, Syria, Taiwan, Tajikistan, Tatarstan, Togo, Turkmenistan, UAE, and Uzbekistan.

As many countries follow the Saudi Arabia lunar crescent visibility criterion, their early criterion is based on Greenwich time zone. Saudi Arabia's old lunar crescent visibility criterion is simply conditioned as: The new Hijri month begins when, after a moon conjunction, sunset occurs before moonset. The old criterion does not consider altitude, age, and elongation.[102] The old criterion is contested by Kordi,[103] as it is not based on the Saudi Arabian time zone or any Saudi Arabia location reference point.

In 2000, a new lunar crescent visibility criterion for Saudi Arabia is introduced. This is in conjunction with the new Umm al-Qurra calendar. The criterion is as follows:

a. The position of lunar crescent and sun is computed using Holy Kaaba as a reference calculation.
b. If a lunar crescent is set before sunset during conjunction, an observation is conducted a day after.
c. If a lunar crescent is set after sunset, its sighting is accepted in accordance with Islamic Jurisprudence of Saudi Arabia.

Mostafa states that the new criterion is based on the capability of a lunar crescent sighting by an observer rather than the capability of a lunar crescent sighting based on a criterion parameter. He adds that this reduces error in lunar crescent report from 14 per cent for old criterion to 0 per cent for the new one. However, as the new Saudi Arabia lunar crescent visibility criterion is solely based on moonset after sunset, it still opens for error in lunar crescent reporting. The world records for lag time are 30 minutes for naked eye, and 20 minutes for optical-aid observations. Should a lunar crescent be observed below the world record limit, then the lunar crescent is highly contestable and should be rejected. However, based on Saudi Arabia's new lunar crescent visibility criterion, a lunar crescent sighting is accepted regardless of its lag time challenge the world records or not.

2.4.4.2 Turkey's Lunar Crescent Visibility Criterion

While Turkey does not have the same magnitude of influence as Saudi Arabia does on Muslim communities worldwide, Turkey has influence in the realm of lunar crescent visibility criterion. Turkey was the first to introduce a lunar crescent visibility criterion in 1978, earlier than Ilyas series of lunar crescent visibility criterion.

Turkey 1978 lunar crescent visibility criterion is the criterion that believed to inspired Malaysia's formation lunar crescent visibility criterion in 1983.[104] Turkey 1978 lunar crescent visibility criterion is a result of an international conference in 1978. The conference was attended by representatives from 20 countries, including Malaysia and Indonesia. The purpose of the conference was to coordinate the determination of new Hijri month among Muslim countries. Through the conference, it has produced several resolutions, among which is a resolution on the Turkey 1978 lunar crescent visibility criterion that is mutually agreed upon by the representative. The criteria that have been agreed upon are as follows:

The new Hijri month began when a lunar crescent

a. Has elongation parameter more than 8 degrees,
b. And has altitude more than 5 degrees.

In 2016, Turkey again proposed another lunar crescent visibility criterion. The proposal was through Conference Islamic calendar in Istanbul Turkey 2016M/1437H.[105] The conference representatives have voted and resolved that

a. The entire world is seen as one union where the new Hijri month begins on the same day throughout that region of the world.
b. A new Hijri month begins when any part of the earth has met the following criteria:
   i. Sun–moon elongation at sunset reaches more than 8 degrees or more.
   ii. Altitude of the lunar crescent is 5 degrees above the horizon.

The criterion, henceforth known as the Istanbul 2016 criterion, acts as a baseline for the International Lunar Dateline. Istanbul 2016 criterion, however seems to ignore a number of lunar crescent observation records.[106] The world records for elongation are 7.7 degrees at naked eye, 6.0 degrees at optical aided, 6.8 degrees at telescopic observation, and 3.42 degrees at CCD imaging. The world records for altitude are 4.06 degrees at naked eye, 6.48 degrees at optical aid, 4.81 degrees at telescopic observation, and 4.62 degrees at CCD imaging.

Indonesia then suggested another criterion in 2017, known as Jakarta Recommendation 2017.[107] The criterion acts as a supplement for the Istanbul 2016 criterion. The resolution on the criterion parameter was conducted through a discussion during the 'International Seminar on Astronomical Fiqh Opportunities and Challenges Implementation of the Single Hijri Calendar', Jakarta on 29–30 November 2017. The conference is attended by participants from five countries, namely Indonesia, Jordan, Malaysia, Singapore, and Brunei Darussalam. Meetings this forms the Jakarta Recommendation 2017. The Jakarta Recommendation 2017 is outlined as:

a. Sun–moon elongation at sunset reaches more or equal to 6.4 degrees.
b. Altitude of the lunar crescent during sunset is more or equal to 3 degrees above the horizon.

Jakarta Recommendation is more suited for Hijri calendar determination as it follows the Islamic Crescent Observation Project world sighting records.

2.4.4.3 MABIMS's Lunar Crescent Visibility Criterion

Malaysia, Indonesia, Brunei, and Singapore employed the lunar crescent visibility criterion to determine their Hijri month. Malaysia, Indonesia, Brunei and Singapore independently determine the first date of *Ramadan* and *Shawal,* at the same time collaborating in formulating lunar crescent visibility criterion for determining the first date of the other Hijri month.[108] In conjunction with the Association of Southeast Asian Nations (ASEAN), these four countries conjoin to form a governing body known as MABIMS (The Informal Meeting of Religious Ministries of Malaysia, Indonesia, Brunei, and Singapore). The role of MABIMS is to conciliate and monitor the laws and principles of lunar crescent sighting and its visibility criterion to reassure there is no disagreement among the members.

As an intercessor for those four countries, each of them is portrayed by their lunar calendar governing bodies; Malaysia is represented by the Department of Islamic Development Malaysia (JAKIM), the Ministry of Religious Affairs Republic Indonesia (KEMENAG RI) for Indonesia, Islamic Religious Council of Singapore (MUIS) for Singapore and the Brunei Islamic Religious Council (MUIB) for Brunei. JAKIM, KEMENAG RI, MUIS, and MUIB are the governing authorities who are responsible for determining the date of the Hijri calendar.

In 1995, Malaysia, Indonesia, Brunei and Singapore have adopted the criterion for lunar crescent visibility, namely '*Imkan al-Rukyah*', which defines the beginning of the lunar months as 'when the lunar crescent could be visible against clear skies'. The concept of this criterion is the 'possibility of visibility' which is based on the result of the visible crescent that has been sighted". By using this lunar crescent visibility criterion, the lunar crescent is respected to be seen when it fulfils one of the following condition:

    a. During sunset
        i. Sun–moon elongation reaches more or equal to 3 degrees.
        ii. Altitude of the lunar crescent is more or equal 2 degrees above the horizon.

Or

    b. During the moonset.
        i. The moon age is more or equal to 8 hours.

The criterion suggests that if a lunar crescent at a certain 29th Hijri day observation has altitude and elongation more than 2 degrees and 3 degrees, respectively, the next day is commenced as new Hijri month. Based on the presence of the lunar crescent in Indonesia from the 1960s to 1990s, the lunar crescent was reported to have appeared several times at an altitude of 2 degrees and elongation of 3 degrees. This criterion was then formulated based on the presence of the lunar crescent, which was confirmed by KEMENAG RI at that time.

The lunar crescent altitude, elongation and moon age parameters that were used in the MABIMS lunar crescent visibility criterion found be to conflicting with other research findings. Elongation criteria have been found by Schafer, Ilyas, Fatoohi and Odeh to be above 7 degrees for naked-eye observation, and none of the lunar crescents are able to be sighted at elongation below 7 degrees, except for extreme optical-aided observation. The same goes to the moon age and lunar crescent altitude, where the world records are 14 hours, and 4 degrees, for moon age and arc of vision, respectively, a parameter that is significantly higher than MABIMS 1995 criterion. This indicates the current lunar crescent visibility criterion adopted by MABIMS outdated, without any current scientific evidence, and is not supported among lunar crescent visibility researchers.

In 2022, Malaysia, Indonesia, Brunei, and Singapore adopt a new criterion for lunar crescent visibility. This criterion is the culmination of research and discussion among researchers, government officials, observatories, and universities in improving the previously flawed 1995 lunar crescent visibility criterion. The new criterion negates the use of the moon age parameter since it has been proven it be ineffective in finding lunar crescent visibility. The new criterion is formed considering the elongation parameter of Odeh lunar crescent visibility criterion, with additional MABIMS's altitude parameter, which originates from the Jakarta Recommendation lunar crescent visibility criterion.

   a. Sun–moon elongation at sunset reaches more or equal to 6.4 degrees.
   b. and altitude of the lunar crescent during sunset is more or equal to 3 degrees above the horizon.

2.4.5 Theme in Designing Lunar Crescent Visibility Criterion

A review was conducted to demonstrate the scientific outlook of a lunar crescent visibility criterion; data locality, prediction strengths and weaknesses, and its long-term legacy in visibility. The reviews demonstrate that each lunar crescent visibility criterion has its own strength, limitation and application for calendrical determination and successful observation. From the review, there are a number of factors that cause the heterogeneity of the lunar crescent visibility criterion.

First are the differences between mathematical models. Caldwell used the lagtime parameter as the main variable in the criteria, showing the existence of the influence of the geocentric model. Modern astronomers are more inclined to mathematical models based topocentric, that is, by using the altitude, azimuth and elongation parameters change according to the position of the observer on the earth's surface. This parameter is a parameter that is used by the majority of experts who study the criteria, such as Ilyas, McNally, and Fotheringham. The difference in terms of the mathematical model used impacts the criterion construction. This is because different mathematical models will use different parameters and subsequently provide heterogeneity in the criterion. This shows the influence of mathematical models on the construction of the criterion.

Next is the primary concept of the criterion designer. Each lunar crescent visibility criterion is built based on one primary concept that is brought by the researcher himself. This concept was born based on the background of society and culture. Thus influences the researcher's thinking and motivation when producing the criteria. A prime example of how a concept can influence a criteria month can be exemplified by Ilyas's lunar crescent visibility criterion. Ilyas is an astronomer atmosphere from Universiti Sains Malaysia. Most of the criteria produced by him are much significantly higher than other criteria of lunar crescent visibility. Researchers such as Fatoohi and Schaefer found the lunar crescent that located above his criterion is easily detected by naked eye. The construction of a high visibility criteria for the moon, which can usually be seen by the human eye, will facilitate the production of calendars for large lines of longitude. The construction of criteria for countries of great longitude is very difficult because the rate of lunar crescent visibility will decrease when going east. This demonstrates that it can be influence on the concept of the Universal Islamic calendar to Ilyas lunar crescent visibility criterion.

Another factor for lunar crescent criterion dissimilarity is the preference on the type of visibility. There are times when the lunar crescent is easily spotted by the naked eye. There are also times when the lunar crescent is vaguely visible and can only be tracked by using a telescope. This results in major complications in the construction of criteria because each researcher has a different visibility preference in building visibility criteria. Some have built criteria based on the condition that it is easily visible by naked eye, and there are also those who construct criteria based on its telescopic visibility. Another example of how selection on the range of visibility has an effect on the results of the lunar crescent visibility criterion is exemplified by This shows that the selection of the range of visibility of the moon, either easily seen with the naked eye or can only be seen using a telescope, affects how a crescent visibility criterion is constructed. An example of this is the difference in approach done by Fotheringham and Maunder in the construction of their criteria. Lunar crescent visibility criterion month expressed by Fotheringham is different from the criterion by Maunder, although both of them used almost the same data in their study. This difference occurs because of the criteria built by Fotheringham based on naked-eye visibility of the lunar crescent. This means, when the parameter of the moon passes the conditions specified by Fotheringham, the crescent moon is definitely visible with naked eyes. Maunder, on the other hand, prefers the critical limit of visibility between positive and negative naked-eye sighting. Therefore, the lunar crescent that located above Maunder limits is easily visible by a telescope but not necessarily visible by naked eye.

The final factor of criteria dissimilarity is the differences in data used for criterion construction. The lunar crescent visibility criterion is built based on empirical data collected by past researchers. There are researchers who have a large collection of lunar crescent sightings; there are also researchers who build their criteria based on a limited collection of lunar crescent sightings. In addition, the amount of lunar crescent sighting data and the distribution of the data influence the graph and criterion of a lunar crescent sighting. An excellent example can be seen in the results of a study conducted by Yallop, Odeh, and Qureshi. Criteria built by Yallop, Odeh, and Qureshi using the same concept, which is the construction of criteria that take into account various ranges of visibility, whether there is visibility with a telescope, visible with the naked eye, and visible with binoculars. They also use the same model, which is the topocentric model, taking into account the width of the crescent moon as the main parameter. However, Yallop, Odeh and Qureshi criteria are different from each other. This is due to the difference in reference data of lunar crescent sightings between Yallop, Odeh and Qureshi. Yallop has 295 data points, Odeh has 737 data points and Qureshi has 436 data points. Differences in data reference and the distribution of the data make the criteria produced by Yallop, Odeh and Qureshi differ from each other even though their criteria are constructed by using the same mathematical concepts and models.

2.5 ASSESSMENT OF LUNAR CRESCENT VISIBILITY CRITERIA

The assessment of lunar crescent visibility criteria is an assessment to measure how dependable the lunar crescent visibility criteria are in predicting the lunar crescent sighting. The assessment can also be used to determine the practicality of a lunar crescent visibility criteria in determining the new Hijri month. Lunar crescent visibility criteria without any assessment cannot be analyzed comparatively as different lunar crescent visibility criteria use different parameters, expressions, and data. Therefore, it is of the utmost importance to assess the lunar crescent visibility criteria. There are few methods to assess the lunar crescent visibility criteria; the methods are lunar crescent sighting records analysis, contradiction rate analysis, qualitative literature analysis, and 29 or 30 cycle analysis.

2.5.1 Lunar Crescent Sighting Records Analysis

Lunar crescent sighting records analysis is an analysis of the lunar crescent visibility criteria based on lunar crescent sighting records. Such example is Fatoohi lunar crescent sighting records analysis on Ilyas elongation criteria. Fatoohi deems that Ilyas 10.5 degree elongation criteria are underestimation of human eye visibility limit, as Fatoohi finds out that there is positive 7.5 degree elongation naked-eye lunar crescent sightings, which are located below Ilyas 10.5 degree elongation limit. Lunar crescent sighting records analysis is able to give an instant assessment of the reliability of the lunar crescent visibility criteria; however, lunar crescent sighting records analysis does not give a full view on the assessment of the lunar crescent visibility criteria as it is based on limited data. This analysis is also found in Schaefer's works on lunar crescent visibility criteria.

2.5.1.1 Literature Analysis

A literature analysis of a lunar crescent visibility criteria is an analysis based on the qualitative assessment of a lunar crescent visibility criteria in comparison to other criteria according to the available literature. A literature analysis enables to demonstrate a glimpse assessment of a lunar crescent visibility criteria reliability and practicality. An example is the assessment of Odeh lunar crescent visibility criteria. As Odeh used the same lunar crescent visibility criteria designed as Yallop, one can draw the similarity, weaknesses, and strengths of Odeh lunar crescent visibility criteria based on Yallop criteria. This will give a glimpse of the performance of an Odeh lunar crescent visibility criterion in predicting lunar crescent visibility. The downside of the criteria is that it does give a proper judgement on a lunar crescent visibility criterion.

2.5.1.2 Lunar Cycle Analysis

Lunar cycle analysis is an analysis of lunar crescent visibility criteria based on how natural the criteria are based on the monthly phase of lunar crescent. The analysis includes a comparison of the lunar crescent visibility criteria on full moon cycle and a comparison of the criteria based on 29- or 30-day Hijri month cycle. An example of lunar cycle analysis is presented by Rodzali and Man, where they assess the frequency of 29th and 30th in over 20 years of Hijri calendar. Another example of lunar cycle analysis is presented by Rahimi and Zainal, where they assess the MABIMS lunar crescent visibility criteria-based accuracy of full month during 15th of a Hijri month, sunset and moonset. The issue with lunar cycle analysis is that the analysis is not justifiable. First, it is not mentioned by the Prophet or any Islamic scholar that the practice of lunar crescent sighting is to ensure the Hijri month follows that lunar cycle accurately; in fact, lunar crescent sighting is practiced during the Prophetic time to cater to the needs and limitations of the companion during his time. Second, the lunar cycle is inconsistent from one lunar cycle to another. Each lunar cycle is unique to one to another depending on the sun and moon declination, position of the earth with respect to the sun and earth rotational speed. Third, there is no practical implication in using lunar cycle to assess the lunar crescent visibility criteria. Assessment of lunar crescent visibility criteria must base on the criteria performance in predicting lunar crescent sighting visibility and its practicality in determining the Hijri calendar. Lunar cycle analysis does not consider a criterion performance in predicting lunar crescent sighting and practicality for Hijri calendar. Considering these three reasons, lunar cycle analysis is not suitable as an assessment method for lunar crescent visibility criteria.

2.5.1.3 Contradiction Rate Analysis

Contradiction rate analysis is an extension of lunar crescent sighting records analysis. While lunar crescent sighting records analysis is based on limited data, contradiction rate analysis is an analysis of lunar crescent visibility criteria based on the whole collection of lunar crescent sighting database. Contradiction rate analysis enables to demonstrate the reliability and practicality of a lunar crescent visibility criteria, as it is based on a large amount of data. This means that it is able to demonstrate the criteria performance at various lunar crescent geometrical parameters and geographical locations. The contradiction rate analysis is divided into two: Positive contradiction rate analysis and negative contradiction rate analysis. Positive contradiction rate analysis is an analysis where the lunar crescent is predicted by the criteria to be sighted but invisible in actual observation. The positive contradiction rate is calculated through the formula of

$$\frac{Total\ Positive\ Moon\ Sighting\ Data - Number\ of\ Positive\ Sighting\ Error}{Total\ Positive\ Moon\ Sighting\ Data} \times 100 \qquad 2.6$$

Negative contradiction rate analysis is an analysis where the lunar crescent is predicted by the criteria to be not sighted but visible in actual observation. The negative contradiction rate is calculated through the formula of

$$\frac{Total\ Negative\ Moon\ Sighting\ Data - Number\ of\ Negative\ Sighting\ Error}{Total\ Negative\ Moon\ Sighting\ Data} \times 100 \qquad 2.7$$

Fatoohi are responsible for introducing the contradiction rate analysis in lunar crescent visibility criteria. Before Fatoohi works on lunar crescent visibility criteria, the analysis of lunar crescent visibility criteria is conducted through limited lunar crescent sighting data analysis. The weakness of the contradiction rate analysis is that it requires large amount of data. Fatoohi amassed a total of 507 data of lunar crescent sighting to conduct his contradiction rate analysis. In addition, the number of positive and negative lunar crescent sighting must be balanced or at least in ratio of 1–3 to ensure a fair analysis on both positive and negative contradiction rate.

2.5.1.4 Histogram Bias Analysis

Histogram bias analysis is an analysis of lunar crescent visibility criteria and contradiction rate in the form of histogram. The histograms use y-axis as the contradiction rate and x-axis as the lunar crescent visibility parameter. The histogram analysis enables us to study the contradiction rate of lunar crescent sighting throughout the increment of the parameter. It highlights the point of the lunar crescent visibility criteria weaknesses.

Histogram bias analysis is introduced by Schaefer (1996) through his paper 'Lunar Crescent Visibility'. He argues that any lunar crescent visibility criteria must have less than 50 per cent bias error in order to be dependable. Lunar crescent visibility criteria that have a large amount of bias in turn have a large parameter of uncertainty, making it not practical for calendrical purposes. Schaefer histogram bias analysis is, however, only applicable for zero-order or singular parameters of lunar crescent visibility criteria. It is not applicable for multi-parameter of lunar crescent visibility criteria. As modern lunar crescent visibility criteria are composed of multi-parameter, histogram bias analysis is not applicable for assessment of lunar crescent visibility criteria.

2.5.1.5 Literature Review on the Assessment of Lunar Crescent Visibility Research

There are number of literatures that conduct the assessment of the lunar crescent visibility criteria. The first to assess the lunar crescent visibility criteria is Ilyas. Ilyas works on lunar crescent visibility criteria are concluded with his conclusive research of the lunar crescent visibility study, which was published in 1994 entitled 'Lunar Crescent Visibility and Islamic Calendar'. In this assessment, Ilyas uses the literature analysis method to analyze the work of various lunar crescent visibility criterion by comparing with his published literature on lunar crescent visibility. Ilyas qualitative assessment of lunar crescent visibility has the preference to favour his own interpretation on lunar crescent visibility criteria, as the literature assessment is heavily based on his literature work and not supported by any lunar crescent sighting data.

The following years saw another assessment of lunar crescent visibility criteria, this time by Schaefer. Schaefer published an assessment lunar crescent visibility criterion in 1996, entitled 'Lunar Crescent Visibility' and with another as a co-author in 1994 with the same title. Schaefer assessment is based on five Moonwatch projects, a nationwide lunar crescent sighting project, taking over 2,000 participants, 294 data points of lunar crescent sighting all across North America longitude. The substantial number of data enables Schaefer to assess various factors that contribute to lunar crescent visibility, which include atmospheric factors, optical factors and human factors. In assessing lunar crescent visibility criteria, Schafer adopts the method of histogram bias analysis, where he assesses the reliability of moon age, lag time, and differences in altitude. However, Schaefer assessment is only limited to these three parameters, while a number of other lunar crescent visibility criteria available during his time are not assessed.

The works of lunar crescent visibility criteria saw another continuation in 1998 in the form of PhD thesis. Louay Fatoohi in 1998 published a PhD thesis where he assessed the lunar crescent visibility criteria based on 506 collections of lunar crescent sighting reports from ancient, medieval, and modern times. Fatoohi uses a combination of lunar crescent sighting records analysis, literature analysis and contradiction rate analysis to assess lunar crescent visibility criteria. This makes his assessment of criteria the most comprehensive assessment even to this day. Fatoohi assessed 15 lunar crescent visibility criteria, including ancient lunar crescent visibility criteria, which are Babylonian and Hindus criteria; medieval lunar crescent visibility criteria which are Khawarizmi, al-Qallas, al-Lathiqi, al-Sanjufini, Ibn Yunus and Maimonides criteria and modern lunar crescent visibility criteria, which are Danjon, Fotheringham, Maunder, Bruin, Ilyas, and Yallop criteria. The only minor disadvantages of Fatoohi lunar crescent visibility criteria are time, as his lunar crescent sighting data collection and criteria assessment are only limited to those available during his time. In the present date, lunar crescent sighting data can be collected over 5,000 data points of lunar crescent sighting, and there are numerous lunar crescent visibility criteria that sprouted after 1998. Fatoohi comprehensive method to assess the lunar crescent visibility criteria should be re-emulated with more database of lunar crescent sightings and modern lunar crescent visibility criteria.

After Ilyas, Schaefer and Fatoohi assessment work on lunar crescent visibility criteria, dedicated work for the assessment of lunar crescent visibility criteria is limited. The current available literature that assesses the lunar crescent visibility criteria are the works of Rodzali and Rahimi, but the literature only assessess single lunar crescent visibility criteria and based on lunar cycle analysis methodology. Thus, there is a need to assess the lunar crescent visibility criteria with updated number of lunar crescent sighting databases and using an articulate analysis methodology.

2.6 CONCLUSION

The discussion on the parameter and criterion of the lunar crescent criterion has demonstrated that each parameter and criterion has its own strengths and weaknesses. Altitude and azimuth are dependent on latitude and prone to error at higher latitudes. Similarly, cases with lag time and moon age parameters are easily calculated, making it a simple rule for determining lunar crescent visibility. Elongation is most effective in determining lunar crescent visibility since it is proven to be related to the length of the lunar crescent.

The discussion on the assessment of lunar crescent visibility criterion shows that the best method to assess the criterion is the combination of lunar crescent sighting records analysis, literature analysis, and contradiction rate analysis. Lunar cycle analysis does not have its impracticality impact, while histogram bias analysis only caters for single parameter type criterion. The discussion also highlights that while the most comprehensive assessment on lunar crescent visibility criterion is Fatoohi's work, the assessment is outdated since it was performed in 1998. Since its publication, numerous reports of recorded lunar crescent sightings and publications of lunar crescent visibility criteria that need to be assessed so that they can be compared analytically with one another.

NOTES

1. Rolf Krauss, "Babylonian Crescent Observation and Ptolemaic-Roman Lunar Dates," *PalArch's Journal of Archaeology of Egypt/Egyptology* 9, no. 5 (2012).
2. Hoffman, "Back to Back Crescent Moon, *The Observatory*, 129 (1208): 1–5.
3. ""The Future of World Religions: Population Growth Projections, 2010-2050," Pew Research Center, 2015, https://www.pewresearch.org/religion/2015/04/02/religious-projections-2010-2050/.
4. Bradley Schaefer, "Lunar Crescent Visibility," *Quarterly Journal of Royal Astronomical Society* (1996): 11.
5. Rolf Krauss, "Babylonian Crescent Observation and Ptolemaic-roman Lunar Dates," *PalArch's Journal of Archaeology of Egypt/Egyptology* 9, no. 5 (2012).
6. Mohammad Ilyas, "Ancients' Criterion of Earliest Visibility of the Lunar Crescent- How Good Is It?" (paper presented at the Proceedings of an International Astronomical Union Colloquium, New Delhi, India, 1987), 4.
7. M. Ilyas, "Limb Shortening and the Limiting Elongation for the Lunar Crescent's Visiblity," *Quarterly Journal of Royal Astronomy Society* 25 (1984): 421–22.
8. Rita Gautschy, "On the Babylonian Sighting-Criterion for the Lunar Crescent and its Implications for Egyptian Lunar Data," *Journal for the History of Astronomy* 45, no. 1 (2014), https://doi.org/10.1177/002182861404500105.
9. Y. Loewinger, "Comments on Bradley Schaefer 1988," *Quarterly Journal of Royal Astronomical Society* 36 (1995): 449–52.
10. J. K. Fotheringham, "The Visibility of the Lunar Crescent," *The Observatory* (1921): 308–11.

11. M. Ilyas, "Age as a Criterion of Moon's Earliest Visibility," *The Observatory* (1983): 26–29.
12. Muhamad Syazwan Faid et al., "Assessment and Review of Modern Lunar Crescent Visibility Criterion," *Icarus* 412 (2024), https://doi.org/10.1016/j.icarus.2024.115970.
13. Andre-Loius Danjon, "Le croissant lunaire [The Lunar Crescent]," *l'Astronomie: Bulletin de la Société Astronomique de France* (1936): 57–65.
14. Nazhatulshima Ahmad et al., "Analysis Data of the 22 Years of Observations on the Young Crescent Moon at Telok Kemang Observatory in Relation to the Imkanur Rukyah Criteria 1995," *Sains Malaysiana* 51, no. 10 (2022), https://doi.org/10.17576/jsm-2022-5110-24; Nazhatulshima Ahmad et al., "A New Crescent Moon Visibility Criteria using Circular Regression Model: A Case Study of Teluk Kemang, Malaysia," *Sains Malaysiana* 49, no. 4 (2020), https://doi.org/10.17576/jsm-2020-4904-15.
15. John Caldwell, "Moonset Lag with Arc of Light Predicts Crescent Visibility," *Monthly Notes of the Astronomical Society of South Africa* 220–235 (2011).
16. J. K. Fotheringham, "On The Smallest Visible Phase of the Moon," *Monthly Notices of the Royal Astronomical Society* 70 (1910): 527–27.
17. Bradley Schaefer, "Lunar Crescent Visibility," *Quarterly Journal of Royal Astronomical Society* (1996): 13.
18. M. Ilyas, "Limb Shortening and the Limiting Elongation for the Lunar Crescent's Visiblity," *Quarterly Journal of Royal Astronomy Society* 25 (1984): 2; Mohd Saiful Anwar Mohd Nawawi, "Penilaian Semula Kriteria Kenampakan Anak Bulan di Malaysia, Indonesia dan Brunei," 2014 Universiti Malaya: Kuala Lumpur; J. K. Fotheringham, "On The Smallest Visible Phase of the Moon," *Monthly Notices of the Royal Astronomical Society* 70 (1910): 3; E. Walter Maunder, "On the Smallest Visible Phase of the Moon," *The Journal of the British Astronomical Association* 21 (1911): 1; Rolf Krauss, "Babylonian Crescent Observation and Ptolemaic-Roman Lunar Dates," *PalArch's Journal of Archaeology of Egypt/Egyptology* 9, no. 5 (2012): 81.
19. Mohammad Odeh, "New Criterion for Lunar Crescent Visibility," *Experimental Astronomy* 18 (2004), https://doi.org/10.1007/s10686-005-9002-5; Muhammad Shahid Qureshi, "A New Criterion For Earliest Visibility of New Lunar Crescent," *Sindh University Research Journal (Sci. Ser.)* 42, no. 1 (2010); B. D. Yallop, *A Method for Predicting the First Sighting of the Crescent Moon*, Nautical Almanac Office (Cambridge: Nautical Almanac Office, 1998).
20. Rita Gautschy, "On the Babylonian Sighting-Criterion for the Lunar Crescent and its Implications for Egyptian Lunar Data," *Journal for the History of Astronomy* 45, no. 1 (2014).
21. Frans Bruin, "The First Visibility of the Lunar Crescent," *Vistas in Astronomy* 21 (1977), https://doi.org/10.1016/0083-6656(77)90021-6, https://linkinghub.elsevier.com/retrieve/pii/0083665677900216.
22. D. L. O'Leary, *How Greek Science Passed to the Arabs* (London: Routledge and Kegan Paul, 1948); R. Morelon, "General Survey of Arabic Astronomy," in *Encyclopedia of the Arabic Science,* ed. R. Rashed (London: Routledge, 1996), 1–19.
23. S. E. Kennedy, "A Survey of Islamic Astronomical Tables," *Transactions of the American Philosophical Society* 46 (1956): 123–77.

24. David A. King, "Some Early Tables for Determining Lunar Crescent Visibility," *Anuals of the New York Academy of Science*, 1987, 185–225.
25. E. S. Kennedy, "Tariq, The Lunar Visibility Theory of Ya'qub Ibn," *Journal of Near Eastern Studies* 27, no. 2 (1986): 126–32.
26. E. S. Kennedy, "The Crescent Visibility Theory of Thabit Bin Qurra," *Proceedings of the Mathematical and Physical Society of the United Arab Republic*, 1960, 71–74.
27. S. E. Kennedy, "A Survey of Islamic Astronomical Tables," *Transactions of the American Philosophical Society* 46 (1956): 123–77.
28. David A. King, "Ibn Yunus On Lunar Crescent Visibility," *Journal of History of Astronomy*, 19, no. 3 (1988): 155.
29. Hamid-Reza Giahi Yazdi, "Al-Khāzinī's Complex Tables for Determining Lunar Crescent Visibility," *Suhayl. International Journal for the History of the Exact and Natural Sciences in Islamic Civilisation* 9 (2009): 149–184,.
30. David A. King, "Some Early Tables for Determining Lunar Crescent Visibility," *Anuals of the New York Academy of Science* 500 (1987): 185–225.
31. Hamid-Reza Giahi Yazdi, "Al-Khāzinī's Complex Tables for Determining Lunar Crescent Visibility," *Suhayl. International Journal for the History of the Exact and Natural Sciences in Islamic Civilisation* 9 (2009).
32. David A. King, "Some Early Tables for Determining Lunar Crescent Visibility," *Anuals of the New York Academy of Science*, 1987, 185–225.
33. Louay Fatoohi, "First Visibility of the Lunar Crescent and other Problems in Historical Astronomy" (PhD, University of Durham, 1998), 51.
34. David A. King, *Astronomy in the service of Islam* (London: Routledge, 1993).
35. M. Ilyas, *Sistem kalendar islam dari perspektif astronomi [The Islamic Calendar System from an Astronomical Perspective]* (Kuala Lumpur: Dewan Bahasa dan Pustaka, 1997).
36. Ebrahim Moosa, "Shaykh Aḥmad Shākir and the Adoption of a Scientifically-Based Lunar Calendar," *Islamic Law and Society* 5, no. 1 (1998): 57–89.
37. Mohd Saiful Anwar Mohd Nawawi , "Penilaian Semula Kriteria Kenampakan Anak Bulan di Malaysia, Indonesia dan Brunei," 2014 Universiti Malaya: Kuala Lumpur,151.
38. J. K. Fotheringham, "On The Smallest Visible Phase of the Moon," *Monthly Notices of the Royal Astronomical Society* 70 (1910): 527–27.
39. Ilyas M. The Ancients' Criterion of Earliest Visibility of the Lunar Crescent: How Good is it? *International Astronomical Union Colloquium*. 1987;91:147-152. doi:10.1017/S0252921100105974
40. August Mommsen, *Chronologie. Untersuchungen über das Kalenderwesen der Griechen insonderheit der Athener* (Leipzig: H.G. Teubner, 1883); J. F. Julius Schmidt, "Ueber die früheste Sichtbarkeit der Mondsichel am Abendhimmel," *Astronomische Nachrichten* (1868), https://doi.org/10.1002/asna.18680711303.
41. Bradley Schaefer, "Atmospheric Extinction Effects on Stellar Alignments," *Journal for the History of Astronomy* 17, no. 10 (1986), https://doi.org/10.1177/002182868601701003, http://journals.sagepub.com/doi/10.1177/002182868601701003.
42. E. Walter Maunder, "On the Smallest Visible Phase of the Moon," *The Journal of the British Astronomical Association* 21 (1911): 355–62.

43. Louay Fatoohi, "First Visibility of the Lunar Crescent and Other Problems in Historical Astronomy" (PhD, University of Durham, 1998), 25.
44. Louay Fatoohi, "First Visibility of the Lunar Crescent and Other Problems in Historical Astronomy" (PhD, University of Durham, 1998), 91.
45. B. D. Yallop, "A Note on the Prediction of the Dates of First Visibility of the New Crescent Moon," in *Astronomical Information Sheet* (1998).
46. John A. Eddy, "The Maunder Minimum," *Science* 192, no. 4245 (1976): 1189–202.
47. Rolf Krauss, "Babylonian Crescent Observation and Ptolemaic-Roman Lunar Dates," *PalArch's Journal of Archaeology of Egypt/Egyptology* 9, no. 5 (2012).
48. Louay Fatoohi, F. Richard Stephenson, and Shetha S. Al-Dargazelli, "The Danjon Limit of First Visibility of the Lunar Crescent," The Observatory, no. 65–72 (1998): 12.
49. Paul Victor Neugebauer, *Astronomische Chronologie* (Berlin: Walter de Gruyter & Co., 1929); Carl Schoch, "The Earliest Visible Phase of the Moon," *The Classical Quarterly* 15, no. 194 (1921).
50. Andre-Loius Danjon, "Le croissant lunaire [The Lunar Crescent]," *l'Astronomie: Bulletin de la Société Astronomique de France* (1936): 57–65.
51. D. McNally, "The Length of the Lunar Crescent," *Quarterly Journal of Royal Astronomy Society* (1983): 417–29.
52. Bradley Schaefer, "Length of the Lunar Crescent," *Quarterly Journal of Royal Astronomical Society* 29 (1991): 511–23.
53. Don Spain, "Apennine Mountains," in *The Six-Inch Lunar Atlas* (Springer, 2009).
54. L. Doggett, Bradley Schaefer, and L. Doggett, "Lunar Crescent Visibility," *Icarus* 107, no. 2 (1994): 388–403.
55. M. Ilyas, "The Danjon Limit of Lunar Visibility: A Re-Examination," *Journal of Royal Astronomy Society Canada* 77, no. 4 (1983): 5.
56. A. H. Sultan, "First Visibility of The Lunar Crescent: Beyond Danjon's Limit," *The Observatory* 53–59 (2007).
57. A. H. Sultan, "New Explanation for Length Shortening of The New Crescent Moon" (2004).
58. Amir Hasanzadeh, "Study of Danjon Limit in Moon Crescent Sighting," *Astrophysics and Space Science* 339, no. 2 (2012), https://doi.org/10.1007/s10509 -012-1004-y, http://www.springerlink.com/index/10.1007/s10509-012-1004-y.
59. David A. King, "Frans Bruin (1922–2001)," *Journal for the History of Astronomy* (2002), https://doi.org/10.1177/002182860203300210.
60. M. J. Koomen et al., "Measurements of the Brightness of the Twilight Sky," *Journal of the Optical Society of America* 42, no. 5 (1952): 353–56.
61. A. Bemporad, "La Teoria della Estinzione Atmosferica nella Ipotesi di un Decrescimento Uniforme della Temperatura dell'Aria coll'Altezza," *Memorie della Societa Degli Spettroscopisti Italiani* 33 (1904): 31–37.
62. H. Siedentopf, "New Measurements on the Visual Contrast Threshold," *Astronomische Nachrichten* 271 (1940): 193–203.
63. S. O. Kastner, "Calculation of The Twilight Visibility Function of Near-Sun Objects," *The Journal of the Royal Astronomical Society of Canada* 70, no. 4 (1976): 153–68.

64. Muhamad Syazwan Faid et al., "Semi Empirical Modelling of Light Polluted Twilight Sky Brightness," *Jurnal Fizik Malaysia* 39, no. 2 (2018): 30059–67.
65. H. R. Blackwell, "Contrast Thresholds of the Human Eye," *Journal of the Optical Society of America* 36, no. 11 (1946), https://doi.org/10.1364/JOSA.40.000825.
66. Louay Fatoohi, "First Visibility of the Lunar Crescent and Other Problems in Historical Astronomy" (PhD, University of Durham, 1998), 121.
67. Muhamad Syazwan Faid et al., "Assessment and Review of Modern Lunar Crescent Visibility Criterion," *Icarus* 412 (2024).
68. M. Ilyas, *Sistem kalendar islam dari perspektif astronomi [The Islamic Calendar System from an Astronomical Perspective]* (Kuala Lumpur: Dewan Bahasa dan Pustaka, 1997), 103.
69. E. Walter Maunder, "On the Smallest Visible Phase of the Moon," *The Journal of the British Astronomical Association* 21 (1911): 5.
70. Frans Bruin, "The First Visibility of the Lunar Crescent," *Vistas in Astronomy* 21 (1977): 21.
71. Louay Fatoohi, "First Visibility of the Lunar Crescent and Other Problems in Historical Astronomy" (PhD, University of Durham, 1998), 110.
72. M. A. McPartlan, "Astronomical Calculation of New Crescent Visibility," *Quarterly Journal of Royal Astronomical Society* (1996): 837–42.
73. Bradley Schaefer, "An Algorithm for Predicting the Visibility of the Lunar Crescent," *Bulletin of the American Astronomical Society* 19 (1987): 1042.
74. Bradley Schaefer, "Visibility of the Lunar Crescent," *Quarterly Journal of Royal Astronomical Society* 29 (1988): 511–23.
75. Leroy Doggett and Bradley Schaefer, "Result of the July Moonwatch," in *Sky & Telescope* (Florida: Sky Publishing, 1989); L. E. Doggett, P. K. Seidelmann, and B. E. Schaefer, "Moonwatch - July 14, 1988," *Sky and Telescope* 76 (1988/07/1988): 34–35.
76. Kevin Krisciunas and Bradley Schaefer, "A Model of The Brightness of Moonlight," *Publications of the Astronomical Society of the Pacific* 103 (1991): 1033–39.
77. Bradley Schaefer, "Heliacal Rise Phenomena," *Journal for the History of Astronomy, Archaeoastronomy Supplement* 18 (1987): 19.
78. Bradley Schaefer, "Atmospheric Extinction Effects on Stellar Alignments," *Journal for the History of Astronomy* 17, no. 10 (1986): 7.
79. Bradley Schaefer, "Telescopic Limiting Magnitudes," *Publications of the Astronomical Society of the Pacific* 102, no. February (1990), https://doi.org/10.1086/132629.
80. Bradley Schaefer, "Visibility Logarithm," in *Sky & Telescope* (Cambridge: Sky Publishing, 1998).
81. Bradley Schaefer, "Lunar Visibility and the Crucifixion," *Quarterly Journal of Royal Astronomical Society* 31, no. 53–67 (1990).
82. Bradley Schaefer, "Astronomy and the Limits of Vision," *Vistas in Astronomy* 36 (1993): 311–61.
83. Bradley Schaefer, "New Methods and Techniques for Historical Astronomy and Archaeoastronomy," *Archaeoastronomy* (2000): 121–21.
84. Louay Fatoohi, "First Visibility of the Lunar Crescent and Other Problems in Historical Astronomy" (PhD, University of Durham, 1998), 37

85. Mohammad Ilyas, *Astronomy of Islamic Times for the Twenty-First Century* (UNKNO, 1988).

86. B. D. Yallop, *A Method for Predicting the First Sighting of the Crescent Moon* (Cambridge: Nautical Almanac Office, 1998).

87. Y. Loewinger, "Comments on Bradley Schaefer 1988," *Quarterly Journal of Royal Astronomical Society* 36 (1995): 449–52.

88. A. H. Sultan, "New Explanation for Length Shortening of The New Crescent Moon," 13.

89. Muhamad Syazwan Faid et al., "Confirmation Methodology for a Lunar Crescent Sighting Report," *New Astronomy* 103 (2023), https://doi.org/10.1016/j.newast .2023.102063.

90. B. D. Yallop and C. Y. Hohenkerk, *A Note on Sunrise, Sunset and Twilight Times and on the Illumination Conditions During Twilight* (1996).

91. Mohammaddin Abdul Niri et al., "The Knowledge Integration Perspective on the Issue of Determining the Time for the Beginning of Fajr Prayer," *Jurnal Fiqh* 16, no. 2 (2019).

92. Paul Victor Neugebauer, *Astronomische Chronologie* (Berlin: Walter de Gruyter & Co., 1929), 111.

93. L. Doggett, Bradley Schaefer, and L. Doggett, "Lunar Crescent Visibility," *Icarus* 107, no. 2 (1994): 14.

94. Louay Fatoohi, "First Visibility of the Lunar Crescent and Other Problems in Historical Astronomy" (PhD, University of Durham, 1998), 74.

95. Muhammad Shahid Qureshi, "A New Criterion For Earliest Visibility of New Lunar Crescent," *Sindh University Research Journal (Sci. Ser.)* 42, no. 1 (2010): 1–16; A. H. Sultan, "'Best Time' For the First Visibility of the Lunar Crescent," *The Observatory* (2006): 115–18.

96. M. Ilyas, "Lunar Calendars: The Missing DateLines," *The Journal of the Royal Astronomical Society of Canada* 80 (1986): 328–35.

97. "Computational Astronomy and The Earliest Visibility of Lunar Crescent," International Crescent Observation Project, updated 2005, http://www.icopro-ject.org/paper.html.

98. John Caldwell, "Moonset Lag with Arc of Light Predicts Crescent Visibility," *Monthly Notes of the Astronomical Society of South Africa* 220–235 (2011).

99. Rita Gautschy and Johannes Thomann, "Dating Historical Arabic Observations," *Proceedings of the International Astronomical Union* 14, no. A30 (2020), https://doi.org/10.1017/s1743921319003983; Rita Gautschy et al., "A New Astronomically Based Chronological Model for the Egyptian Old Kingdom," *Journal of Egyptian History* 10, no. 2 (2017), https://doi.org/10.1163/18741665 -12340035.

100. Rita Gautschy, "On the Babylonian Sighting-Criterion for the Lunar Crescent and its Implications for Egyptian Lunar Data," *Journal for the History of Astronomy* 45, no. 1 (2014): 4.

101. T. Alrefay et al., "Analysis of Observations of Earliest Visibility of the Lunar Crescent," *The Observatory* 138, no. 1267 (2018).

102. Zaki Mostafa, "Lunar Calendars: The New Saudi Arabian Criterion," *The Observatory* 125 (2005): 25–30.

103. Ayman Kordi, "The Psychological Effect On Sighting of The New Moon," *Observatory* (2003): 2.

104. Mohd Saiful Anwar Mohd Nawawi, Saadan Man, Mohd Zambri Zainuddin, Raihana Abdul Wahab, and Nurulhuda Ahmad Zaki, "Sejarah Kriteria Kenampakan Anak Bulan Di Malaysia," *Journal of Al-Tamaddun* 10, no. 2 (2015/12/31): 332.

105. Susiknan Azhari, "Cabaran Kalendar Islam Global di Era Revolusi Industri 4.0," *Jurnal Fiqh* 18, no. 1 (2021/6/2021), https://doi.org/10.22452/fiqh.vol18no1.4, https://fiqh.um.edu.my/index.php/fiqh/article/view/30691.

106. Abdul Mufid and Thomas Djamaluddin, "The Implementation of New Minister of Religion of Brunei, Indonesia, Malaysia, and Singapore Criteria Towards the Hijri Calendar Unification," *HTS Teologiese Studies / Theological Studies* 79, no. 1 (2023), https://doi.org/10.4102/hts.v79i1.8774; Abdul Mufid, "Unification of Global Hijrah Calendar In Indonesia: An Effort To Preserve The Maqasid Sunnah of The Prophet (SAW)," *Journal of Islamic Thought and Civilization* 10, no. 2 (2020).

107. Maskufa Maskufa et al., "Implementation of the New MABIMS Crescent Visibility Criteria: Efforts to Unite the Hijriyah Calendar in the Southeast Asian Region," *AHKAM: Jurnal Ilmu Syariah* 22, no. 1 (2022), https://doi.org/10.15408/ajis.v22i1.22275.

108. Mohd Saiful Anwar Mohd Nawawi et al., "Pemikiran Imam Taqī Al-Dīn Al-Subkī (683/1284-756/1355) Berkaitan Kriteria Kenampakan Anak Bulan," *Jurnal Syariah* 28, no. 1 (2020).

Development of an Analysis Tool in Python for Lunar Crescent Sighting and Criteria

3

3.1 INTRODUCTION

This chapter discusses the development of an analysis tool for the lunar crescent visibility criteria using an integrated lunar crescent observation database. This chapter is divided into four parts: Data collection, data calculation, data analysis, and tool development. Data collection details the data mining methodology in collecting the lunar crescent observation database from literature. Data calculation details the recalculation of the data using the latest astrometry library, Skyfield. Next, data analysis details the Python library used to analyze the data and the types of the analysis that can be conducted on the data. Methodology chapter concludes with the design and development of a Python package for the analysis tool for the lunar crescent visibility criteria

DOI: 10.1201/9781003536192-3

using integrated lunar crescent sighting database and the steps in utilizing the Python library.

3.2 DATA COLLECTION

This section entails the data collection methodology for the lunar crescent sighting database. This section is divided into four parts. The first part is data source, a subsection that reveals the literature source for the new lunar crescent sighting database. The second part is data mining, a subsection that discusses the methodology of data mining from the literature and website, combining text-based mining methodology, image-based mining methodology, and web-based mining technology. The third part is data cleaning, the removal of the arguable reports in the database. The last part is information addition, which includes the addition of the time zone and elevation.

3.2.1 Data Source

Eleven literatures are identified as reliable databases of lunar crescent sighting reports, which are Yallop, Fatoohi, Hoffman, Odeh, Qureshi, Caldwell, Krauss, Hasanzadeh, Musfiroh and Hendri, Rehman et al., and Alrefay et al.[1] Cumulatively, these literatures amass a total amount of 3,962 data points of lunar crescent sightings. Data of lunar crescent sightings from Islamic Crescent Observation Project website from 1998 to 2020 is also scrapped.[2] In total, 8,290 data points of lunar crescent sightings are collected for this work.

3.2.2 Data Mining

3.2.2.1 Text-Based Lunar Crescent Sighting Data Extraction

Text-based literature is literature that can contain extractable text and can be edited using PDF editor tools. The text mining methodology of this literature can be done by analyzing and detecting the table structure of a lunar crescent sighting report. Tables that are structured with clear boundary lines are lattice structures, while tables that are separated by invisible tabs or spaces are defined as stream structures. The structure of the table, either lattice or stream, is identified using edge detection. Edge detection able to recognize the header of the table and the boundary of the data from one column or row to another.

Edge detection then converts the boundary and separation between two values in the table into comma-separated value (CSV) data. These data can be formatted into an excel file and enable the user to modify the structure and analyze the data. These steps are operated using the Python library Camelot. Camelot is a Python library that functions to detect tables in text-based PDF format and convert the table into a CSV file.[3]

3.2.2.2 Image-Based Lunar Crescent Sighting Data Extraction Flowchart

Image-based literature is the literature and documents portrayed are usually exemplified by scanned document or documents written with old typing technology, such as typewriters. Since format of these texts does not follow the modern design of computer text writing, Camelot unable to detect the text in the document, which consequently renders its tabular detection ability useless.

Detecting a table in an image-based document similarly uses an edge detection technique, with contrast difference, where it detects the table boundary using image detection rather than text detection. The boundary is defined by recognizing the text and non-text from image detection. The non-text image is interpreted as an edge, which acts as columns or rows from one table to another. These steps are operated using the Python library ExtractTable. ExtractTable is a Python library to detect tables in image-based documents and convert the table in CSV file.[4] ExtractTable uses artificial intelligence to detect tabular data in images and has the flexibility to edit the data.

3.2.2.3 Web-Based Lunar Crescent Sighting Data Extraction

Web-based data are the data that are stored on the website. The format of website data is different from one website to another, thus scraping the data requires understanding the website structure. Scraping data from a website can be performed using Beautiful Soup, a Python library.[5] Beautiful Soup is a library that specializes in web scraping. For data on lunar crescent sighting, the Islamic Crescent Observation Project and Moonsighting website are scraped to collect the records of lunar crescent sightings.

ICOP website data of lunar crescent sightings are text data. This means that the data is not stored in a tabular format but in text format. In the text, the website highlights the date of the sighting, location of the sighting, time of the sighting, status of the sighting, condition of the sky during the sighting, methodology of the sighting and a photographic proof. The developer mode of the ICOP website able to provide a more structured presentation where the elements of 'date of the sighting', 'location of the sighting', 'time of the sighting', and 'status of the sighting' are separated into different classes. This can help in scraping the data of lunar crescent observations. Scraping the data of lunar crescent sightings from ICOP website is operated using the Beautiful Soup Python Library. Beautiful Soup can recognize the website structure and extract the class that contains information about lunar crescent sighting data. The extracted data is then stored in a data frame format to enable manipulation of the data.

The columns 'raw text' contain the information about 'location of the sighting', 'time of the sighting', 'status of the sighting', and 'methodology of the sighting'. This information is presented in a structured format, enabling it to be extracted using a programming extraction technique called regular expression. Regular expression, or shortened as regex, is a character that is sequenced to help in a search pattern. String extraction for the method of sighting is utilized to identify the tools that used by the observer to locate the lunar crescent. The extraction is possible because the ICOP lunar crescent sighting methodology is systematically arranged into the following order:

the crescent was seen by naked eye, the crescent was not sought by binocular, the crescent was not sought by telescope, the crescent was not sought by CCD Imaging

The order demonstrated that lunar crescent visibility information is determined by two items, which are sighting 'was seen' or 'was not sought', indicating lunar crescent visibility and invisibility. The methods of sighting include 'by naked eye', 'by binocular', 'by telescope', and 'by Charge-Coupled Device-CCD) imaging', which indicate sighting by naked eye, sighting by binocular, sighting by telescope and sighting by CCD Imaging. This information can be extracted by identifying the text 'was seen' and its subsequent sentences, thereby extracting the information on the lunar crescent visibility and its method of sighting.

The extracted text requires further cleaning. First, the absence method of sighting indicates that the lunar crescent is not sighted with that method of sighting. Second, the lunar crescent must have the keyword 'was seen by' to denote a positive sighting; otherwise, it would be considered a negative sighting. String extraction for witness is utilized to locate the name of the witness who reported the lunar crescent visibility. The extraction of possible witness naming is systematically arranged into the following order:

Time of observation. After sunset. Not Seen. ICOP member Mr. Abdellatif Fareh from City in State mentioned that the sky was clear, the atmospheric condition was hazy, the crescent was not sought by naked eye, the crescent was not sought by binocular, the crescent was not sought by telescope, the crescent was not sought by CCD Imaging

The order demonstrated that the name of the witness was sandwiched between the term 'ICOP member' and 'from City'. Therefore, a text extraction could be done by extracting the text after the term 'ICOP member' and splitting the text before the term 'from City'. The extraction code is as follows:

```
regex = 'ICOP .*\w'
df['test'] = df['Raw_Text'].str.extract('('+regex+')',
expand=True)
witness = df["test"].str.split("from ", n = 1, expand =
True)
df["witness"]= witness[0]
df.drop(columns =["test"], inplace = True)
```

String extraction for sky condition is utilized to measure the condition of the sky during lunar crescent observation. The extraction is possible as sky condition is systematically arranged into the following order:

Time of observation. After sunset. Not Seen. ICOP member Mr. Abdellatif Fareh from City in State mentioned that the *sky was clear*, the atmospheric condition was hazy, the crescent was not sought by naked eye, the crescent was not sought by binocular, the crescent was not sought by telescope, the crescent was not sought by CCD Imaging.

The order demonstrated that the condition of the sky is sandwiched between the term 'sky' and a comma. Therefore, a text extraction could be done by extracting the text after the term 'sky' and splitting the text before the comma. The extraction code is as follows:

```
#sky condition
regex = 'the sky .*\w'
df['test'] = df['Raw_Text'].str.extract('('+regex+')',
expand=True)
sky = df["test"].str.split(",", n = 1, expand = True)
df["Cloud"]= sky[0]
df.drop(columns =["test"], inplace = True)
```

String extraction for sky condition is utilized to measure the condition of the sky during lunar crescent observation. The extraction is possible as the sky condition is systematically arranged into the following order:

> Time of observation. After sunset. Not Seen. ICOP member Mr. Abdellatif Fareh from City in State mentioned that the sky was clear, the atmospheric condition **was hazy**, the crescent was not sought by naked eye, the crescent was not sought by binocular, the crescent was not sought by telescope, the crescent was not sought by CCD Imaging

The order demonstrated that the condition of the sky is sandwiched between the term 'atmospheric condition' and a comma. Therefore, a text extraction could be done by extracting the text after the term 'atmospheric condition' and splitting the text before the comma. The extraction code is as follows:

```
#Atmospheric Condition
regex = 'the atmospheric .*\w'
df['test'] = df['Raw_Text'].str.extract('('+regex+')',
expand=True)
atmospheric = df["test"].str.split(",", n = 1, expand =
True)
df["Atmospheric Condition"]= atmospheric[0]
df.drop(columns =["test"], inplace = True)
```

ICOP only presents the location of the lunar crescent sighting in the form of name of the location. The website does not provide the latitude and longitude coordinates of the location. The latitude and longitude are integral variables of lunar crescent sighting to recalculate the geometrical position of the lunar crescent during observation. The location coordinate extraction is operated using the Python library. Geopy uses openstreetmap.com data to identify the location coordinates (latitude and longitude). Geopy also able to read location data in Arabic, which is the majority of the presented names of the location in ICOP website.

3.2.3 Data Filtering

3.2.3.1 Data Frame of Lunar Crescent Sighting Data Conversion and Filtering

Data frames are Python format of data in tabular form. The data frame is only readable in Python; thus, it needs to be converted into a readable data format, which is CSV format. Then, the converted CSV file requires filtering, as some of the data contains information that is not relevant for lunar crescent sighting data analysis. Filtering of the data is executed manually to ensure the validation of the data extraction using Python.

3.2.4 Data Validation

The data is validated based on the information from the original sources. If the original source is unable to confirm the data, or the source contradicts itself, the data will be eliminated.

> Time of observation: After sunset. Not Seen: ICOP member Dr. Abdurrazak Ebrahim from Cape Town City in Western Province State mentioned that the sky was clear. the atmospheric condition was clear, the crescent was seen by the naked eye, the crescent was not sought by binocular, the crescent was not sought by telescope, the crescent was not sought by CCD Imaging.

The instance of unvalidated data: The data above reported that the lunar crescent is not sighted; however, the following statement reported that the lunar crescent is observed through the naked eye. This is contradictory data and needs to be eliminated from the database.

3.3 DATA CALCULATION

This section will entail the data calculation methodology for the lunar crescent sighting database. This section will be divided into three parts. The first part is data structure, a subsection that reveals the structure for data on lunar crescent sightings. Second part is ephemeris calculation, a subsection that discusses the calculation of the parameters essential for lunar crescent sightings. The third part is data authentication, which means the removal of the inaccurate reports in the data.

3.3.1 Data Structure

The data of lunar crescent sighting is structured based on data presentation by Odeh,[6] which contains reference number (Ref No), year (Y), month (M), day (D), latitude (Lat), longitude (Long), and time of observation (O); with (E) for evening observation, and (M) for morning observation, visibility of lunar crescent sighting (V); with (V) for visible, and (I) for invisible, methodology of sighting (M); with (NE) for naked eye, and (OA) for optical aided, imaging technique (I); with (NU) for not used, (CCD) for charge-couple device observation and (T) for Digital Single Light Reflect (DSLR) telescopic observation, time zone (TZ), moon age (MA), lag time (LT), arc of vision (ArcV), arc of light (ArcL), difference in azimuth (DAZ), and width of the lunar crescent in seconds (W).

Latitude, longitude, and date are core variables for lunar crescent sighting data. Julian day number is included to further validate the date of the lunar crescent observation to avoid error. Julian day number is calculated using Meeus algorithm.[7] Arc of vision, arc of light, difference in azimuth, and width are the most preferred variables by researchers for lunar crescent visibility criterion, thus included in the data structure. Visibility of lunar crescent sighting indicates whether a sighting is positive or negative. Contrary to most literature practice where negative sighting is not included, the author decided to include negative sighting to enable the positive and negative sighting assessment of any lunar crescent visibility criterion. The reference number consists of the source of the lunar crescent sighting report and the order number of the report according to the original. The reference number is shortened, and list of the shortened references is as shown in Table 3.1.

3.3.2 Ephemeris Coding

3.3.2.1 Python Library

The ephemeris is calculated using the Skyfield Python library. Skyfield is an updated Python library for PyEphem, used for computing the positions of sun, moon, stars, and other celestial objects. The Skyfield Python library is a high-calibre astrometry library that can generate celestial object positions with the precision of 0.0005 arcseconds, in agreement with the United States Naval Observatory (USNO) and their Astronomical Almanac.[8]

3.3.2.2 Date and Time

The lunar crescent data include observations from the Ancient Era, Middle Ages, and numerous data from ancient times. The data year ranges from 566 BC to 1185 AD. These data are dated prior to the event of the Julian-Gregorian Date conversion, 15 October 1582, where 13 days were omitted from the Julian Calendar, from 4 October 1582 to 15 October 1582. The calendar, called the Gregorian Calendar, omitted 13 days to accommodate the dating of Christian Easter Time. This causes the dating of the lunar crescent visibility records for

TABLE 3.1   Data Referencing and Source

REFERENCE	SOURCE
Fatoohi	Louay Fatoohi, 'First Visibility of the Lunar Crescent and Other Problems in Historical Astronomy' Durham University, 1998
Qur	Muhammad Shahid Qureshi, 'A New Criterion for Earliest Visibility of New Lunar Crescent', *Sindh University Research Journal (Sci. Ser.)*, 42.1 (2010): 1–16.
Yallop	B.D. Yallop, A Method for Predicting the First Sighting of the Crescent Moon, Nautical Almanac Office (Cambridge. Nautical Almanac Office, 1998).
ICOP	Islamic Crescent Observation Project
Ho	Roy E. Hoffman, 'Back to Back Crescent Moon', *The Observatory*, 129, no. 1208 (2009): 1–5.
ALR	T. Alrefay and others, 'Analysis of Observations of Earliest Visibility of the Lunar Crescent', *The Observatory*, 138, no. 1267 (2018).
Amir	Amir Hasanzadeh, 'Study of Danjon Limit in Moon Crescent Sighting', *Astrophysics* *and Space Science*, 339.2 (2012), 211–21 <https.//doi.org/10.1007/s10509-012-1004-y>.

Krauss	Rolf Krauss,
	'Babylonian Crescent Observation and Ptolemaic-Roman Lunar
	  Dates',
	PalArch's Journal of Archaeology of Egypt/Egyptology, 9.5
	  (2012).

Ca	John Caldwell, 'Moonset Lag with Arc of Light Predicts Crescent Visibility,' *Monthly Notes of the Astronomical Society of South Africa*, 220–235 (2011).

Mus	Imas Musfiroh and Hendri, 'Analisis Regresi Non Linier
	  (Polynomial) Dalam Pembentukan Kriteria Visibilitas Hilal Di
	  Indonesia',
	Al-Marshad. Jurnal Astronomi Islam Dan Ilmu-Ilmu Berkaitan,q
	  2018 <https.//doi.org/10.30596/jam.v4i1.1935>.

TK	Malaysia, Teluk Kemang Observatory Data

Source: Researcher Data.

the Ancient and Middle Ages to be off by 13 days. Skyfield has a feature to correct this issue, which is simply include this line into the coding.

```
from skyfield.api import GREGORIAN_START
ts.julian_calendar_cutoff = GREGORIAN_START
```

This code will perform Julian-Gregory Date conversion and enable the calculation of moon and sun ephemeris for ancient times to be more precise. Some of the more popular lunar ephemeris software such as Accurate Times, Moon Calc, and Neoprogramics are unable to calculate the negative years due to this issue. Fortunately, by using Skyfield, this issue is negated.

3.3.2.3 Calculation of Sunset

Skyfield can find sunset using its almanac.rising_setting API module. The module requires the data of latitude, longitude, year, month and day. The result will return the time of the sunset.

3.3.2.4 Calculation of Moonset

The moonset calculation using Skyfield is operated as same as sunset calcula-tion. However, there are instances where moonset does not occur on the same day as sunset. This occurs when sunset at a location is near the 24-hour mark, such as 11:30 p.m. or 11:45 p.m. This will result in an error in code execution. This can be negated by extending the range from one day to two days. There is also the case where moonset occurs earlier than the sunset. This happens because the data does not include time zone.

3.3.2.5 Calculation of Lag Time

Skyfield does not include the calculation of lag time in its library. Thus, the lag time needs to be calculated manually. Coding-wise, however, the sunset and moonset are presented in the form of a string or text, thus cannot be computed. This can be solved by converting the hour and minute into float individually, and then the lag time formula can be performed. The lag time in the table is presented in minutes.

3.3.2.6 Calculation of Moon Age

Similarly, with lag time, moon age calculation is not included in the Skyfield library. Thus, it needs to be calculation manually. The moon age is defined by the age of the moon from the conjunction to the time it is sighted; in this case, it is assumed that the moon is sighted at sunset. Moon age is presented in hours. Coding-wise, moon age is calculated by subtracting the Julian date number at sunset from the sunset to Julian date number during moon conjunction. The code execution is as below.

3.3.2.7 Calculation of Julian Day Number

Julian day number is a presentation of the calendar in a day format. The Julian day number starts from 4716 BC and extends to infinite year of AD. Julian day number enables the standardization of dates across the Gregorian, Julian, Muslim, or Hebrew calendars. Skyfield includes Julian day numbers in its API library.

3.3.2.8 Calculation of Arc of Vision

Arc of vision is defined as the vertical angle between the moon's altitude and the sun's altitude. Skyfield Python library does not include arc of vision in its API; thus, it needs to be calculated manually.

3.3.2.9 Calculation of Arc of Light

Arc of light is defined as the angle between the moon's altitude and the sun's altitude. Skyfield Python library does include arc of light in its API. Skyfield interprets arc of light or elongation as the degree of separation, which in this API is referred as x.separation_from(y).format.

3.3.2.10 Calculation of Different in Azimuth

Different in zimuth, or DAZ as the name suggests, is defined as the angle between moon azimuth and sun azimuth. The Skyfield Python library does not include difference in azimuth in its API; thus, it needs to be calculated manually.

3.3.2.11 Calculation of Width

Width is defined as the illuminated angle of the moon. In this thesis, it is expressed in terms of arc second. Skyfield Python library does not include width in its API. To determine width, the Jean Meeus and Yallop algorithm is used.

TABLE 3.2   ICOP Records of Lunar Crescent Observation

CATEGORY	PARAMETER	VALUE
Naked-eye observation	Moon age	15 hours 1 minute
	Lag time	29 minutes
	Elongation	7.7 degrees
Optical-aid observation	Moon age	11 hours 17 minutes
	Lag time	20 minutes
	Elongation	6.0 degrees
Ordinary imaging	Moon age	12 hours 40 minutes
	Lag time	20 minutes
	Elongation	6.8 degrees
CCD	Moon age	0 hours 0 minutes
	Elongation	3.42 degrees

Source: ICOP.

3.3.3 Data Authentication

The data from the source need to be authenticated to ensure the reliability of the data. Data authentication is conducted by using the records' visibility limits as listed on the ICOP website. The parameters of the visibility records are demonstrated in Table 3.2.

Based on the parameters in Table 3.2, data that is located below parameter is removed from consideration. This is because the ICOP database has the most extensive compilation of lunar crescent sightings, making its records more dependable than others.

3.4 DATA ANALYSIS

Data analysis discusses the analysis conducted on the data.

3.4.1 Analysis Based on Limiting Threshold

The threshold analysis is important to understand the limits of lunar crescent visibility based on two methods of sighting, which are naked-eye sighting and optical aid. The analysis enables researchers to examine the threshold parameters in ensuring the sighting of the moon. The threshold analysis can be examined using a box graph. The box graph is chosen because it provides a summarized view of the median, lower, and upper thresholds, and first and third quartiles of the data.

From the box plot, information about the position of the minimum and maximum values, the first and third quartiles, and the location of the median of the data is provided. The maximum and minimum values enable the reader to understand the range of the data. In the case of lunar crescent sighting, minimum value is used as a record of lunar crescent observation and a tool for data validation. The first quartile, third quartile and median tell the reader the position of the data. In the case of lunar crescent sighting, data located above the first quartile parameter could be said to have a 25 per cent chance of sighting, data located above the median parameter could be said to have a 50 per cent chance of sighting, and data located above the third quartile parameter could be said to have a 75 per cent chance of sighting. Box plot information, as shown in Figure 3.1, combined with standard deviation and mean, can provide ample information about the pattern of data in the graph. The figure shows that A is the location of the minimum, B is the location of the first quartile of the data, C is the median, D is the location of the third quartile of the data, and E is the location of the maximum.

Source: Researcher Data.

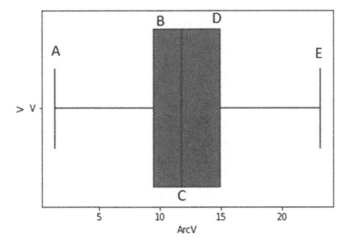

FIGURE 3.1   Instance of a Boxplot.

3.4.2 Analysis Based on Condition

An instance of a conditional style lunar crescent visibility criterion is the MABIMS 2021 lunar crescent visibility criterion. The conditional analysis is able to analyze the percentage of accuracy in predicting the visibility of the lunar crescent or the contradiction rate of positive sightings and, subsequently, contradiction rate of negative sightings. The dataset of the lunar crescent visibility report is then portrayed using a scatter plot. A scatter plot is chosen because it is the best graph to represent the two important classes of the data, method of observation and positive or negative sighting. A line graph is also overlaid on top of the scatter plot to represent the conditional parameter of MABIMS lunar crescent sightings.

The same code execution is conducted on naked eye, binocular, telescope, and CCD sighting records. The dataset is then evaluated with the condition parameter of MABIMS lunar crescent sightings. The test result is divided into two categories: Positive sighting contradiction rate and negative contradiction rate. Any positive lunar crescent visibility data that is below the parameter condition is counted as positive sighting contradiction rate. Any unsuccessful lunar crescent visibility data that is above the parameter condition is counted as negative sighting contradiction rate.

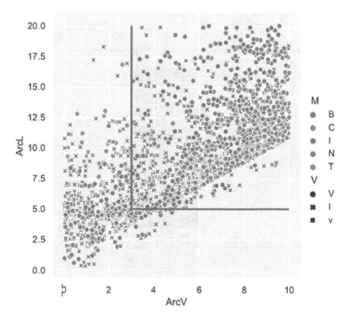

FIGURE 3.2   Conditional Analysis Sample. Source: Researcher Data.

3.4.3 Analysis Based on Equation

An instance of equation type lunar crescent visibility criterion is Fatoohi lunar crescent visibility criterion. The equation analysis is able to analyze the percentage of accuracy in predicting the visibility of the lunar crescent or the contradiction rate of positive sighting, and the contradiction rate of negative sighting.

The user is required to insert the parameter being analyzed, the parameter condition, and the limiting database for the lunar crescent visibility. Fatoohi's lunar crescent visibility criterion is set as an example for this purpose. The dataset of the lunar crescent visibility report is then portrayed using a scatter plot. A functional graph from the equation is overlaid on top of the scatter plot to represent the equation parameter of Fatoohi's lunar crescent sighting criterion. The equation is read by using the *eval* function.

The same code execution is conducted on naked eye, binocular, telescope and CCD lunar crescent sighting records. The database of lunar crescent sighting is then evaluated with the equation parameter of the Fatoohi criterion. The

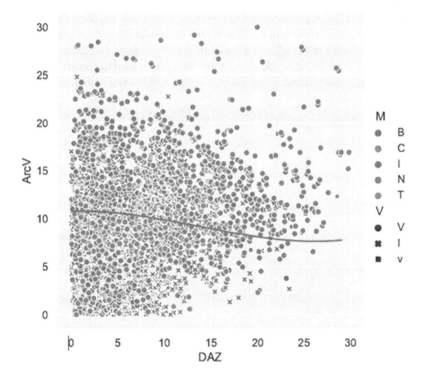

FIGURE 3.3   Result of Equation Analysis. Source: Researcher Data.

test result is divided into two categories: Positive sighting contradiction rate and negative sighting contradiction rate. First, a new column is created based on the value of the equation from Fatoohi using numexpr, a numerical expression evaluator. Then, the parameter is evaluated based on the value from the equation.

3.4.4 Analysis Based on Regression Analysis

Analyzing lunar crescent visibility criteria involves understanding the accuracy and predictive power of a criterion for lunar crescent visibility. Three key statistical measures, which are mean absolute error (MAE), mean square error (MSE), and R-squared (R^2), are used to evaluate these criteria. This is portrayed in Figure 3.4.

MAE measures the average magnitude of errors in a set of predictions without considering their direction. It is the average over the test sample of the absolute differences between prediction and actual observation, where all individual differences have equal weight. A lower MAE value indicates a better fit. A lower MAE value indicates that the model's predictions are, on average, very close to the actual observations. This means the model is highly accurate in its predictions, with minimal deviation from the true values. For lunar crescent visibility, a lower MAE would suggest that the criteria used to predict visibility are effective and reliable. Conversely, a higher MAE value indicates that the

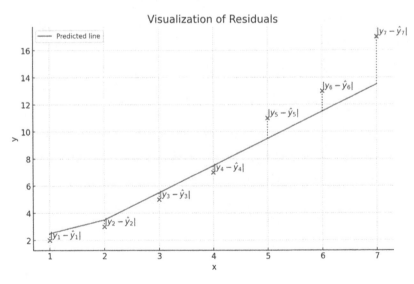

FIGURE 3.4   Instance of Mean Absolute Error.

predictions are, on average, farther from the actual observations, suggesting less accuracy. This would imply that the predictive model needs improvement or that the criteria may not be robust enough.

MSE measures the average of the squares of the errors, that is, the average squared difference between the estimated values and the actual values. Unlike MAE, MSE gives more weight to larger errors. A lower MSE value indicates a better fit, though it is more sensitive to outliers than MAE. A lower MSE value indicates that the model's predictions are, on average, very close to the actual observations, with minimal large errors. This suggests that the model is highly accurate and reliable in its predictions. Conversely, a higher MSE value indicates that the predictions are, on average, farther from the actual observations, suggesting less accuracy. Larger errors are penalized more heavily, indicating potential outliers or significant deviations that the model fails to predict accurately. R^2, also known as the coefficient of determination, indicates how well data fits a statistical model. It provides the proportion of the variance in the dependent variable that is predictable from the independent variable(s). An R^2 value closer to 1 indicates that the model explains most of the variance in the actual visibility observations. This signifies a good fit, meaning that the model's predictions closely match the actual data. High R^2 values suggest that the independent variables (predicted values) are good predictors of the dependent variable (actual visibility observations). Conversely, an R^2 value closer to 0 indicates that the model explains very little of the variance in the actual visibility observations. This signifies a poor fit, meaning that the model's predictions do not closely match the actual data. Low R^2 values suggest that the independent variables are not good predictors of the dependent variable.

Using a combination of MSE, MAE, and R^2 is crucial for a comprehensive evaluation of predictive models for lunar crescent visibility. Each metric provides unique insights into different aspects of model performance, ensuring a well-rounded assessment.

MAE measures the average magnitude of errors between predicted and actual visibility values, offering a straightforward indication of how close the model's predictions are to the actual observations. It is particularly useful for understanding the typical size of errors without giving undue weight to large outliers. The simplicity and interpretability of MAE make it an easy metric to understand, as it uses the same units as the original data. This allows researchers to quickly grasp the average deviation from actual visibility observations, helping to identify if the model consistently overestimates or underestimates visibility.

MSE measures the average of the squared differences between predicted and actual values, providing more weight to larger errors. This sensitivity to larger errors helps identify models that may have occasional significant errors, which could be critical in lunar crescent visibility predictions where large deviations might significantly impact decision-making. By squaring the errors, MSE emphasizes outliers more than MAE, helping researchers understand the impact of extreme values on the model's performance. This can be particularly important for refining models to reduce the occurrence of large prediction errors.

R^2, also known as the coefficient of determination, indicates the proportion of the variance in actual visibility observations that is explained by the model. It provides a measure of how well the model captures the underlying data patterns. A high R^2 value suggests that the model is effective in explaining the variability in lunar crescent visibility, which is crucial for reliable predictions. R^2 helps in assessing the overall fit of the model, showing how well the predicted values correspond to the actual data, giving a sense of the model's effectiveness in capturing the relationships between variables. It is essential for determining whether the model is robust and generalizable to different datasets.

Using all three metrics together provides a balanced perspective on model performance. MAE gives an average error magnitude, MSE highlights large errors and R^2 assesses the explanatory power of the model. Together, they offer a more nuanced understanding of the model's strengths and weaknesses, helping researchers make informed decisions about model improvements. Analyzing MSE, MAE, and R^2 together helps identify specific areas where the model may need improvement, such as reducing large errors or improving overall fit. This comprehensive evaluation ensures that the model is not only accurate on average (MAE) but also robust against outliers (MSE) and effective in explaining the variability in the data (R^2).

For lunar crescent visibility predictions, precision and reliability are crucial. Combining these metrics ensures that the model is both accurate and reliable, providing confidence in its predictions for practical applications like determining the start of months in lunar calendars. Using MAE, MSE, and R^2 together ensures a thorough evaluation of predictive models for lunar crescent visibility, ensuring that the models meet the critical requirements for precise lunar observations.

The regression analysis is conducted based on the minimum data of the lunar crescent visibility criterion. The parameter of a lunar crescent visibility criterion is selected. The x-axis of the parameter is set as a fixed variable, with a determined interval. The y-axis of the parameter is based on the minimum value of the parameter, derived from its corresponding value on the x-axis parameter.

```
df = pd.DataFrame(data_mongo)
#####
df[a] = df[a].abs()
df[b] = df[b].abs()
```

```
#print(dataset)
dftest = df[df['V'] =='V']
if dataset == "Whole":
dftest=dftest
else:
dftest = dftest[dftest['M'] ==dataset]
dftest = dftest[[a,b,'V','M','I']]
df_test_sorted = dftest.sort_values(a)
df_test_sorted = df_test_sorted[df_test_sorted[a] <=
limita]
df_test_sorted = df_test_sorted[df_test_sorted[b] <=
limitb]
# Create a Selected DataFrame
df = df_test_sorted
min_range = 1
max_range = limita
increment = 1
results = []
for i in range(min_range, max_range + 1, increment):
mask = (df[a] >= i) & (df[a] < i + increment)
subset = df[mask]
if len(subset) > 0:
min_b = subset.groupby(by='M')[b].min()
for m, min_b_value in min_b.items():
results.append('x': f'i + increment - 1', 'y': min_b_
value, 'M': m)
result_df = pd.DataFrame(results)
df = result_df
df['x'] = df['x'].astype(np.int64)
```

After the selection, the regression analysis on the lunar crescent visibility criterion is conducted. The regression analysis is conducted using the scikit-learn Python library, for MAE, MSE, and R^2. The coding for the regression analysis, based on various equations of lunar crescent visibility criterion, is

```
x = df['x'].astype(float).values.reshape(-1, 1)
y_pred1 = eval(equation1)
mae1 = mean_absolute_error(df['y'], y_pred1.ravel())
mse1 = mean_squared_error(df['y'], y_pred1.ravel())
r21 = r2_score(df['y'], y_pred1.ravel())
# Plot the actual values against the predicted values
plt.figure(figsize=(20,10))
chart=sns.relplot(x=df['x'], y=df['y'],hue=df['M'])
plt.plot(x, y_pred1, color='red', label=f'criterion_1')
plt.xlabel(a)
plt.ylabel(b)
```

```
plt.legend()
plt.show()
chart.figure.savefig("RegressionAssesment.p
ng",bbox_inches='tight')
document.add_picture("RegressionAssesment.png")
paragraph = document.add_paragraph("7Figure 7: Graph of
the Criterion over Minimum Value of Lunar Crescent Data")
df1 = {'Criterion': ['Mean Absolute Error (MAE)','Mean
Squared Error (MSE)','R Squared'],
f'{criterion_1}': [mae1,mse1,r21],
```

3.5 ANALYSIS TOOL DEVELOPMENT

Library development describes the design and development of a Python package for the analysis tool for the lunar crescent visibility criteria using integrated lunar crescent sighting database and the steps in utilizing the Python library.

3.5.1 Hilalpy[9]

Development analysis tool for the lunar crescent visibility criteria based on the integrated lunar crescent sighting database is the purpose of this work. The analysis tool is called HilalPy, due to its applicability in lunar crescent or Hilal visibility research and its development in a Python programming language (ascl:2307.031). HilalPy is packaged as a Python library and contains three functions, which are:

- To summarize the descriptive statistics of lunar crescent visibility parameters based on the integrated lunar crescent sighting database.
- To provide an analysis of the contradiction rate of the lunar crescent visibility criteria based on the integrated lunar crescent sighting database.
- To provide analysis of distance regression of lunar crescent visibility criteria based on the integrated lunar crescent sighting database.

Hilalpy is available at the website https://pypi.org/project/hilalpy/. The collected database contains 8,290 records of lunar crescent visibility, 3,023 negative records of lunar crescent sightings, and 5,267 positive records of lunar crescent sightings. Out of 5,267 positive records of lunar crescent sightings, 4,092 are naked-eye lunar crescent sightings, and 1,175 are optical-aided lunar crescent sightings. Out of 1,175 optical-aided lunar crescent observations, 335 are captured digitally using CCD, 191 are captured digitally using telescopes, and 648 data of the lunar crescent are observed without any digital imaging. Out of 3,023 negative records of lunar crescent sightings, 3,013 records are naked-eye observations, and 10 records are optical-aided observations.

3.5.2 Hilal-Obs[10]

The confirmation methodology is developed using the Python library Skyfield for the astrometric function (ascl:1907.024). Skyfield is chosen because the result computation agrees with the United States Naval Observatory and their Astronomical Almanac to within 0.0005 arcseconds in computing the topocentric position of the sun and moon. The code is registered in the Astrophysics Source Code Library as HilalObs (ascl.soft04003F). To use the method, users should copy the coding into their Python integrated development environment. The coding requires Skyfield to work and works best within the Anaconda Environment. Therefore, it is advisable to download Anaconda and the Skyfield Library. Further instructions on using the code are available at the HilalObs Github Repository.[11]

For example, if the user wants to validate a lunar crescent report named 'Tested Data' during the evening of 1 May 2022, at 5.90 North Latitude, 116.04 East Longitude, within the GMT +8:00 time zone, the user should conduct the following steps.

1. Determine the level of light pollution at the observation site. This can be obtained from direct measurement using a sky quality meter or by using a public light pollution map website. For the coordinates of 5.90 North Latitude, 116.04 East Longitude, it has 20.30 mag/sec^2 as reading of light pollution.

2. Locate elevation in metres, temperature in Celsius, and humidity in decimal during observation. For example, 5.90 North Latitude, 116.04 East Longitude, has an elevation of 6 m, a temperature of 29°C, and humidity of 0.77.

3. Create CSV file with nine columns and two rows, containing Reference No (Ref No), which is the name of the data; Latitude (Lat), positive for northern latitude, negative for southern latitude; Longitude (Long), positive for eastern longitude, negative for western longitude; Day; Month; Year; Reading of Light Pollution in mag/sec^2 (LP); Time Zone (TZ); Type of Observation, 'E' for evening, 'M' for morning (O), Ele for elevation in metres, Temp

for temperature in Celsius, and Humidity for humidity in decimal points for the first row; and lunar crescent data for the second row.

4. Run HilalObs coding on the CSV file, with the updated file location. HilalObs coding produces another two columns, which are the Modified Schaefer Prediction Model and the Crumey Prediction Model. If a lunar crescent was computed as sighted, the 'Moon Sighted' output appears, while if the lunar crescent was computed as not sighted, an NaN value appears in the second column.

5. For the tested data, Modified Schaefer found that the lunar crescent was computed to be sighted, while the Crumey model computed the lunar crescent to be not sighted.

Another example, if the user wants to validate a lunar crescent report named 'Syawal Observation' during the evening of 20 April 2023, at 38.97 North Latitude, 95.23 West Longitude, with a GMT −5:00 time zone, the user should conduct the following steps:

1. Determine the level of light pollution at the observation site. This can be obtained from direct measurement using a sky quality meter or by using a public light pollution map website. For the coordinates of 38.97 North Latitude, 95.23 West Longitude, it has 22.00 mag/sec^2 as reading of light pollution.

2. Locate elevation in metres, temperature in Celsius, and humidity in decimal during observation. For example, 38.97 North Latitude, 95.23 West Longitude, has an elevation of 252 m, a temperature of 6°C, and humidity of 0.67.

3. Create CSV file with nine columns and two rows, containing Reference No (Ref No), which is name of the data; Latitude (Lat), positive for northern latitude, negative for southern latitude; Longitude (Long), positive for eastern longitude, negative for western longitude; Day; Month; Year; Reading of Light Pollution in mag/sec^2 (LP); Time Zone (TZ); Type of Observation, 'E' for evening, 'M' for morning (O) for the first row; and lunar crescent data for the second row.

4. Run HilalObs coding on the CSV file, with the updated file location. HilalObs coding produces another two columns: Modified Schaefer Prediction Model and Crumey Prediction Model. If a lunar crescent is computed as sighted, the 'Moon Sighted' output appears, while if the lunar crescent is computed as not sighted, an NaN value appears in the second column.

5. For the tested data, both the Modified Schaefer and Crumey models computed the lunar crescent to be not sighted.

The confirmation methodology of lunar crescent visibility records is evaluated to ensure its reliability in predicting lunar crescent visibility in actual observations. The test is conducted by examining the model's capability to predict both visibility and invisibility of 1,869 reports of lunar crescent observations. The accuracy of the modelling test is shown on the Hilal-Obs GitHub site.[12] These data have 15 columns, which are:

Column 1. Ref No: Reference Number of the Lunar crescent visibility report followed by its numbering sequence based on the original source.

Column 2. Lat: Latitude. North is positive and South is negative.

Column 3. Long: Longitude. East is positive and West is negative.

Column 4, 5, and 6. Day, Month, Year: Date of the Lunar Crescent Sighting

Column 7. V: Visibility. V for a positive moon sighting, I for a negative moon sighting.

Column 8. Lag time: Time difference between sunset and moonset in minutes.

Column 9. Moon Age: Age of the Moon from conjunction in hours. Calculated at the exact sunset.

Column 10. ArcV: Arc of Vision. Angle between sun altitude and moon altitude in degrees. Calculated at the exact sunset.

Column 11. ArcL: Arc of Light. Angle between sun and moon in degrees. Calculated at the exact sunset.

Column 12. DAZ: Difference in Azimuth. Angle between sun azimuth and moon azimuth in degrees. Calculated at the exact sunset.

Column 13. Width: Topocentric lunar width in minutes. Calculated at the exact sunset.

Column 14. Schaefer: Modified Schaefer prediction result. I for the moon predicted not to be visible, V for the moon is predicted to be visible.

Column 15. Crumey: Crumey prediction result. I for the moon predicted not to be visible, V for the moon is predicted to be visible.

The test has found that the Crumey prediction model outperforms the Modified Schaefer prediction model by a slight margin. The Modified Schaefer model mispredicted 37 per cent of negative moon sightings, while Crumey mispredicted 28 per cent of negative moon sightings. For predicting positive sightings, the Modified Schaefer model has an error rate of 7.42 per cent, while the Crumey model has an error rate of 7.89 percent.

It is found that the Crumey model, despite not being assessed on any lunar crescent visibility criterion, is capable of predicting moon sightings. The Crumey model is also suitable for light-polluted sky brightness, making the model applicable for the majority of moon sighting observation sites. In addition, lunar crescents are observed before the end of dawn, within the sky brightness range of 14 mag/sec^2 to 20 mag/sec^2, a value similar to the Crumey effective visibility range. The Modified Schaefer model, despite having laudable accuracy, is based on this research experiment's empirical results, while Crumey is supported by both theoretical and empirical works. The Modified Schaefer model is also more reliable in predicting the telescopic visibility of the lunar crescent under an elongation of 6 degrees, while the Crumey model is better at predicting the lunar crescent visibility.

3.6 CONCLUSION

This chapter demonstrates the methodology involved in developing analysis tool for the lunar crescent visibility criteria using integrated lunar crescent sighting database. The chapter demonstrates the method of data collection, including text mining from available literature, web scraping from websites, transformation of the data into data frame format, and cleaning the data from unnecessary information. The chapter then demonstrates the calculation of lunar crescent sighting data using Skyfield and cleaning of unauthenticated data of lunar crescent sightings. The chapter is then followed by an illustration of the method to analyze the lunar crescent visibility criterion, using contradiction rate analysis based on the lunar crescent sighting database and statistical analysis of lunar crescent observation variables. The chapter proceeds with the process of developing HilalPy, a Python library that functions to analyze the lunar crescent visibility criteria based on integrated lunar crescent sighting database, and HilalObs, a Python library to validate the visibility of a lunar crescent.

NOTES

1. T. Alrefay et al., "Analysis of Observations of Earliest Visibility of the Lunar Crescent," *The Observatory* 138, no. 1267 (2018); Louay Fatoohi, "First Visibility of the Lunar Crescent and Other Problems in Historical Astronomy"

(PhD, University of Durham, 1998); Amir Hasanzadeh, "Study of Danjon Limit in Moon Crescent Sighting," *Astrophysics and Space Science* 339, no. 2 (2012): 211–21; Roy E. Hoffman, Observing the new Moon, *Monthly Notices of the Royal Astronomical Society* 340, no. 3 (2003): 1039–51; J.A.R. Caldwell, & C. D. Laney, "First visibility of the lunar crescent." *MNASSA: Monthly Notes of the Astronomical Society of South Africa* 58, no. 11 (1999): 150–63; Rolf Krauss, "Babylonian Crescent Observation and Ptolemaic-Roman Lunar Dates," *PalArch's Journal of Archaeology of Egypt/Egyptology* 9, no. 5 (2012); Musfiroh, Imas. "Analisis regresi non linier (polinomial) dalam pembentukan kriteria visibilitas hilal di indonesia." *Al-Marshad: Jurnal Astronomi Islam dan Ilmu-Ilmu Berkaitan* 4, no. 1 (2018); Mohammad Odeh, "New Criterion for Lunar Crescent Visibility," *Experimental Astronomy* 18 (2004): 39–64; Muhammad Shahid Qureshi, "A New Criterion for Earliest Visibility of New Lunar Crescent," *Sindh University Research Journal (Sci. Ser.)* 42, no. 1 (2010): 1–16; B. D. Yallop, *A Method for Predicting the First Sighting of the Crescent Moon,* Nautical Almanac Office (Cambridge: Nautical Almanac Office, 1998); T. Alrefay et al., "Analysis of Observations of Earliest Visibility of the Lunar Crescent," *The Observatory* 138, no. 1267 (2018); Louay Fatoohi, F. Richard Stephenson, and Shetha S. Al-Dargazelli, "The Danjon Limit of First Visibility of the Lunar Crescent," *The Observatory*, no. 65–72 (1998); Louay Fatoohi, "First Visibility of the Lunar Crescent and Other Problems in Historical Astronomy" (PhD, University of Durham, 1998), 112.
2. https://astronomycenter.net/record.html.
3. Mehta Vinayak, "camelot-py," (Python Package Index, 26/2/2023), https://pypi.org/project/camelot-py/#description.
4. "ExtractTable-py: Python Library to Extract Tabular Data from Images and Scanned PDFs," Github, 2022, https://github.com/ExtractTable/ExtractTable-py.
5. "Beautiful Soup Documentation," Crummy, 2020, https://www.crummy.com/software/BeautifulSoup/bs4/doc/.
6. Mohammad Odeh, "New Criterion for Lunar Crescent Visibility," *Experimental Astronomy* 18 (2004): 39–64.
7. Jean Meeus, *Astronomical Algorithms* (Virginia: Willmann-Bell, 1991).
8. Brandon Rhodes, "Skyfield: High Precision Research-grade Positions for Planets and Earth Satellites Generator" (Astrophysics Source Code Library, 2019), https://ascl.net/1907.024.
9. M.S. Faid et al., "Assessment and Review of Modern Lunar Crescent Visibility Criterion," *Icarus* 412 (2024); Muhamad Syazwan Faid, Mohd Saiful Anwar Mohd Nawawi, and Mohd Hafiz Mohd Saadon, "HilalPy: Analysis Tool for Lunar Crescent Visibility Criterion," (Astrophysics Source Code Library, 2023); Muhamad Syazwan Faid et al., "HilalPy: Software to Analyse Lunar Sighting Criteria," *Software Impacts* 18 (2023), https://doi.org/10.1016/j.simpa.2023.100593; M. S. Faid, M. S. A. Mohd Nawawi, and M. H. Mohd Saadon, "Analysis Tool for Lunar Crescent Visibility Criterion based on Integrated Lunar Crescent Database," *Astronomy and Computing* 45 (2023), https://doi.org/10.1016/j.ascom.2023.100752.

10. M. S. Faid et al., "Confirmation Methodology for a Lunar Crescent Sighting Report," *New Astronomy* 103 (2023): 102063–63, https://doi.org/10.1016/j.newast.2023.102063; Muhamad Syazwan Faid et al., "Hilal-Obs: Authentication Agorithm for New Moon Visibility Report" (Astrophysics Source Code Library, 2021), ascl:2104.003.

11. https://github.com/msyazwanfaid/Authentication-Algorithm-for-New-Moon-Visibility-Report.

12. https://github.com/msyazwanfaid/Authentication-Algorithm-for-New-Moon-Visibility-Report/blob/master/Result%20of%20Calculated%20Contrast.csv.

Assesment of the Lunar Crescent Visibility Criteria

4

4.1 INTRODUCTION

This chapter discusses the assessment of the lunar crescent visibility criterion. It contains four parts: the database of lunar crescent sightings, visibility threshold analysis, contradiction rate analysis, and discussion. The database of lunar crescent sightings subchapter entails a summary of the collected data on lunar crescent sightings. The visibility threshold analysis discusses the box plot analysis of the lunar crescent visibility parameters and its descriptive statistics. The contradiction rate analysis discusses the analysis of the modern lunar crescent visibility criterion.

4.2 DATABASE OF LUNAR CRESCENT OBSERVATION[1]

4.2.1 Database Composition

The collected database contains 8,290 records of lunar crescent visibility; 3,023 records are negative records of lunar crescent sighting, while 5,267 are

DOI: 10.1201/9781003536192-4

positive records of lunar crescent sighting. Out of 5,267 positive records of lunar crescent sighting, 4,092 are naked-eye lunar crescent sightings, and 1,175 are optical-aided lunar crescent sightings. Out of 1,175 optical-aided lunar crescent observations, 335 are captured digitally using CCD, 191 are captured digitally using telescopes, and 648 data of lunar crescent are observed without any digital imaging. Out of 3,023 negative records of lunar crescent sightings, 3,013 records are naked-eye observations, and 10 records are optical-aided observations.

4.2.2 Removed Data of Lunar Crescent Observation

As the data of lunar crescent observation comprises of various sources, there is data that is unvalidated, incorrect, or inconsistent and requires further authentication to ensure the reliability of the data. The method of authentication is discussed in Chapter 4, which includes the following:

a. Data Filtering

Removing the data that contradict or are unconfirmed by their initial source. Using this methodology, 22 data are removed from the database, as shown in Table 4.1.

b. Data Validation

Validation of the data using ICOP lunar crescent sighting records: Data that are located below the ICOP Lunar crescent observation records are removed from the analysis to ensure the reliability of the data. Data from credible sources, however, using this methodology, there are 233 data points that are removed from the analysis using this methodology, as shown in Table 4.2.

TABLE 4.1   Filtered Data of Lunar Crescent Sighting

DATABASE	REASON OF REMOVAL	RECORDS
ICOP	Contradictory data	7
Krauss 2012	Unconfirmed data	15
Total		22

Source: Researcher Data.

TABLE 4.2   Removed Records of Lunar Crescent Sighting Database

DATABASE	REASON OF REMOVAL	RECORDS
Musfirah 2015	70 data or 65 per cent of the data contradicts the ICOP records. Total rejection of the data source due to a sizeable percentage of contradiction.	138
Caldwell 2012	25 data or 26 per cent of the data contradicts the ICOP records. Total rejection of the data source due to a sizeable percentage of contradiction.	95
Fatoohi 1998	15 data contradict the ICOP records.	15
Krauss 2012	9 data contradict the ICOP records	9
Odeh 2005	1 data contradict the ICOP records	1
Hoffman 2004	6 data contradict the ICOP records	6
Teluk Kemang	1 data contradict the ICOP records	1
Total		265

Source: Researcher Data.

Musfiroh and Hendri and Caldwell data are totally removed from the analysis as the majority of their data contradicts the records of lunar crescent by ICOP. The reasoning for Musfiroh and Hendri contradiction is due to unauthenticated records of lunar crescent sightings, which are prevalent issue in Indonesia, particularly records that predate 1995.[2] Caldwell database removal is due to a number of miscalculations of Caldwell's lunar crescent visibility data. For example, Caldwell included data from 7 January 2000. Moon conjunction in January 2000 happened on 24 January 2000. Another example, Caldwell has positive records of lunar crescent during 16 January 1991, with a topocentric elongation of 3.03 degrees and a lag time of 25 minutes. This is far below other world records of lunar crescent sightings. A number of Krauss and Fatoohi also contradicted ICOP world records of sightings. The reason for the contradiction is the unconfirmed date of sightings, an issue that hampers ancient records of lunar crescent observation the most. Krauss and Fatoohi contain numerous records of lunar crescent observation from the Babylonian to Medieval eras. Hoffman and Odeh's contradiction records of lunar crescent sightings are due to contradictions in reporting. The total number of accepted data of lunar crescent observations is 8,025.

4.3 INSTANCE OF DISPUTABLE LUNAR CRESCENT VISIBILITY REPORT[3]

Data 1: Indonesia Data 2022

Shawwal in Malaysia during May 2022 was celebrated on a different precalculated date. Initially, the lunar crescent during the twenty-ninth of Ramadhan, or the first of May, was determined to be undetectable. The twenty-ninth Ramadhan lunar crescent has a topocentric arc of vision of 5.9 degrees and an elongation of 6.13 degrees, which does not pass the new criterion adopted by Malaysia, which is 4 degrees of arc of vision and 6.4 degrees of elongation. However, despite the precalculated lunar crescent being invisible, the lunar crescent was reportedly sighted in Malaysia and Brunei using CCD and was reported to be seen by the naked eye in multiple locations in Indonesia, such as Balai Rukyat Condrodipo Gresik by a team of observers and Tanjung Kodok Lamongan by three observers and Pontianak. The data of the lunar crescent are portrayed in Table 4.3.

All reported sighting data do not pass the parameters of naked-eye sighting as outlined by Fatoohi, Schaefer, Ilyas, and Odeh.[4] Bukit Shahbandar, Balai Rukyat Condrodipo, and Pontianak are located outside urban areas with light pollution estimations of 19.36 mag/sec^2, 21 mag/sec^2, and 19.55 mag/sec^2, respectively. Crumey's modelling revealed that the lunar crescent is not visible

TABLE 4.3 Contestable Report of Lunar Crescent Sighting in Malaysia, Brunei, and Indonesia

			PARAMETER IN TOPOCENTRIC		
DATA	LOCATION	METHOD OF OBSERVATION	ARC OF VISION	ARC OF LIGHT	DIFFERENCE IN AZIMUTH
Malaysia	Balai Cerap Al-Biruni Sabah 5.94 N, 116.05 E	CCD	5°37′	5°42′	0°53′
Brunei	Bukit Shahbandar 4.95 N, 114.86 E	Telescope and naked eye	5°38′	5°43′	1°01′
Indonesia	Balai Rukyat Condrodipo Gresik 7.17 S, 112.62 E	Naked eye	5°13′	5°45′	2°25′
	Pontianak 0.03 S, 109.32 E	Naked eye	5°38′	5°52′	1°40′

to the naked eye. Should these data be sighted by a telescope or CCD, then they are viable data; however, if the lunar crescent was sighted using naked-eye sighting, then they are counted as arguable data of lunar crescent sighting.

Data 2: India 2018

In 2018, Ramadhan was celebrated on two different dates. Ramadhan was celebrated on 18 May by Metro City of Mumbai and its neighbouring cities, while other parts of India celebrated on 17 May. The lunar crescent for the month of Ramadhan was sighted in three southern states of India, Tamil Nadu, Karnataka, and Telangana, on 16 May, and this sighting was accepted by the majority of Islamic groups in India, including Imarat-e Shariah Bihar, Jharkhand, Orissa, Jamiat Ulama-e Hind, and the Central Hilal Committees of Jama Masjid Delhi and Nakhoda Masjid, Kolkata, and Ahle Hadith group in Mumbai. However, the Hilal Committees of Mumbai from both Deobandi and Barelvi schools rejected the sighting and did not consider the testimony to be valid; therefore, they announced that Mumbai would not consider the 16 May lunar crescent sighting and began their Ramadhan on 18 May instead. Data on the lunar crescent are portrayed in Table 4.4.

The lunar crescent parameter on 16 May passed the threshold of the visible naked eye set by Fatoohi, Alrefay, and Odeh (Alrefay et al., 2018; Fatoohi, 1998; Odeh, 2004). Both Karnataka and Tamil Nadu have excellent observation sites with an average light pollution of 21.54 mag/sec^2. Lunar crescent sightings are predicted by Crumey modelling; thus, the decision of the Hilal Committees of Mumbai to reject the 16 May lunar crescent sighting is not entirely validated.

Data 3: Canada 2020

Canada saw various groups of lunar crescent sightings. The Crescent Council of Canada and the Hilal Council of Canada follow international lunar crescent sightings from California, Saudi Arabia, UAE, Malaysia, and Turkey. Another group, the Hilal Committee of Toronto and Vicinity, does not adhere

TABLE 4.4 India's Contestable Lunar Crescent Sighting

DATA	LOCATION	METHOD OF OBSERVATION	ARC OF VISION	ARC OF LIGHT	DIFFERENCE IN AZIMUTH
			PARAMETER IN TOPOCENTRIC		
India	Karnataka 12.53 N, 76.55 E	Naked eye	13°22′	14°05′	4°28′
India	Tamil Nadu 8.13 N, 77.31 E	Naked eye	13°35′	13°59′	3°21′

to any international lunar crescent sightings and follows their own local sightings for new Hijri month determination. In Ramadhan 2020, the Crescent Council of Canada and Hilal Council of Canada announced that Ramadhan in Canada began on 24 April, while the Hilal Committee of Toronto and Vicinity began their Ramadhan on 25 April. Their data on the lunar crescent are portrayed in Table 4.5.

The lunar crescent parameter for both the twenty-second and twenty-third centuries has a high chance of sighting due to the high value of the arc of light. It lies above the limit of the naked eye by Ilyas (1984) and Odeh (2005). However, Crumey's modelling finds that the lunar crescents on the twenty-second and twenty-third are predicted to be invisible. This is due to the high level of pollution, which is 17.87 mag/sec^2 in Toronto. If the lunar crescent be observed at a pristine, light pollution-free site, it has a higher chance of visibility; however, if the lunar crescent observation is conducted in Toronto, then it has a lower chance of visibility.

Data 4: Saudi Arabia 2020

In June 2019, Saudi Arabia sparked controversy over a reported sighting of the lunar crescent at the end of Ramadan on 3 June 2019. The lunar crescent, which has the parameters of 1°23′ for relative sun–moon altitude and 3°40′ for elongation, was located below the visibility line and should not have been sighted by either optical aid or naked-eye observation. However, since Saudi Arabia acts as a Hijri calendar reference, some mosques and Islamic communities in Britain followed Saudi Arabia in this setting. One of the Saudi Arabian location sites, Abha, is a pristine observation site for lunar crescent sightings, with a level of light pollution of 21.61 mag/sec^2. However, it does not help with the extreme parameter of lunar crescent sighting. The lunar crescent sighting on 3 June 2019 was not sighted according to Crumey modelling; thus, the decision to accept the lunar crescent sighting was not validated.

TABLE 4.5   Canada's Contestable Lunar Crescent Sighting

			PARAMETER IN TOPOCENTRIC		
DATE	LOCATION	METHOD OF OBSERVATION	ARC OF VISION	ARC OF LIGHT	DIFFERENCE IN AZIMUTH
22 April 2020	Toronto 43.73 N, 79.30 W	Naked eye	3°40′	14°05′	4°28′
23 April 2020		Naked eye	6°29′	10°02′	7°40′

Data 5: Pakistan 2018

On 14 June 2018, in Pakistan, a lunar crescent was reported to be visible to the naked eye. The lunar crescent was reported to be visible by congregations in Masjid Qasim Ali Khan (Bilal, 2018). The report is arguable because all neighbouring regions and Peshawar said that the lunar crescent was not visible. In addition, Masjid Qasim Ali Khan congregations have a 150-year history of false-positive reports of moon sightings. Masjid Qasim Ali Khan also reports 14 testimonies of positive moon sightings. The contested report led to the divided celebration of Shawwal between the Pakistan Government on 16 June 2018 and Masjid Qasim Ali Khan congregations on 15 June 2018. While being contested, the arguable report of moon sightings has an elongation of 11.7 degrees and an arc of vision of 7.32 degrees, a parameter that passes the threshold of naked-eye visibility by Fatoohi, Alrefay, and Odeh (Alrefay et al., 2018; Fatoohi, 1998; Odeh, 2004). Masjid Qasim is located in the city centre of Peshawar city, with a light pollution estimation of 18.9 mag/sec^2 and an extinction coefficient of 0.45 based on a light pollution map and extinction coefficient by Schaefer (Falchi et al., 2016; Schaefer, 1986). Our modelling found that the lunar crescent is not visible to the naked eye due to the high level of light pollution. Thus, the decision of the Ruet-i-Hilal Committee of the Pakistan Government to reject the observation of Masjid Qasim Ali Khan and its 14 positive observation records is validated.

Data 6: Nigeria 2020

Nigeria had three different commencement dates for Ramadhan during 2020. The Chief Imam of Ibadan Land and Grand Patron of the League of Imams and Alfas of Yorubaland announced that Ramadhan should begin on 23 April 2020, while the President of the Nigerian Supreme Council for Islamic Affairs declared that Ramadhan should be observed on 24 April 2020, and the League of Imams and Alfas of South–West Nigeria stated that Ramadhan should begin on 25 April 2020. Decisions on 23 April and 25 April were based on theoretical data, while those on 24 April were based on actual sightings. Astronomical calculations showed that the Moon had not passed the conjunction phase by 22 April 2020, making the determination of 23 April as the commencement date of Ramadhan outrightly rejected. The modelling results found that the moon could not be sighted on 23 April 2020, making the decision to commence the date of Ramadhan on 24 April 2020 incorrect. At sunset on 23 April 2020, the lunar crescent had a crescent width of 7 arc minutes and an elongation of 7.24, making its size too small to be detected by the naked eye. The lunar crescent was sighted on 24 April 2020, making the decision to start Ramadhan on 25 April 2020 correct. At sunset on 24 April 2020, the lunar crescent had a width of 44.45 arc minutes, which was large enough to be visible to the naked eye.

TABLE 4.6 Results of Contestable Records of Moon Sighting

REF NO	LAT	LONG	DAY	MONTH	YEAR	LIGHT POLLUTION	CRUMEY PREDICTION MODEL
Condrodipo Gresik 2022	−7.17	112.62	1	5	2022	19.36	
Pontianak 2022	−0.03	109.32	1	5	2022	21	
Brunei 2022	4.95	114.86	1	5	2022	19.55	
Karnataka 2018	12.53	76.55	16	5	2018	21.54	Moon sighted
Tamil Nadu 2018	8.13	77.31	16	5	2018	21.27	Moon sighted
Pakistan	34.01	71.57	14	6	2018	18.99	
Canada 22 April 2022	43.73	−79.3	22	4	2020	17.87	
Canada 23 April 2022	43.73	−79.3	23	4	2020	17.87	
Saudi Arabia 2022	18	42	3	6	2019	21.61	
Nigeria (Ibadan)	7.4	3.92	22	4	2020	20.12	
Nigeria (Sokoto)	13.06	5.23	23	4	2020	19.71	
Nigeria (Southwest)	6.58	3.75	24	4	2020	21.42	Moon sighted
South Africa	−26	28	27	7	2014	18.9	Moon sighted
Kenya (Mombasa)	−4.05	39.67	12	5	2021	19.76	Moon sighted

Data 7: South Africa 2014

South Africa also observed a disagreement in determining the date of Shawwal in 2014. The United Ulama Council of South Africa (UUCSA) declared that Shawwal 2014 was to be celebrated on 28 July 2014 due to reports of lunar crescent sightings in Gauteng, North Riding, and Lenasia South. Some Muslims rejected the declaration and began their Shawwal celebration on 29 July 2014. The rejection was due to some Muslims challenging the validity of the lunar crescent sighting on 27 July 2014, citing that the observation was not accepted, as the witness was not from a Sunni background. The model determined that the lunar crescent on 27 July 2014 had a high possibility of being sighted. At sunset, the lunar crescent had a lunar altitude of 7.7 degrees, elongation of 8.2 degrees, width of 9.02 arc minutes, and azimuth difference of 2.81 degrees. These parameters just sat above the visibility line, making the lunar crescent sighting challenging. Despite the arguable nature of the lunar crescent parameters, the Moon was found to be sighted by both Crumey and Modified Schaefer modelling of the contrast threshold.

Data 8: Kenya 2021

Kenyans observed Shawwal on two distinct occasions in 2021. Kenya Chief Kadhi, Ahmed Muhdhar, announced that Ramadhan commenced on 14 May 2021, while another group claimed that the lunar crescent was sighted on 12 May 2021, making Ramadhan end by 12 May and Shawwal begin on 13 May 2021. Modelling has found that the lunar crescent on 12 May 2021 had a high possibility of being sighted. At sunset, the lunar crescent had a lunar altitude of 8.03 degrees, elongation of 8.31 degrees, width of 9.28 arc minutes, and azimuth difference of 2.15 degrees. These parameters lie just above the visibility line, making Ahmed Muhdhar's action to disregard the lunar crescent sighting justified. Despite the contentious nature of the lunar crescent parameter, both Crumey and Modified Schaefer contrast threshold models find the Moon to be visible. The results are shown in Table 4.6.

4.4 RESULT OF VISIBILITY THRESHOLD ANALYSIS[5]

This subsection discusses the threshold limit of lunar crescent visibility on various parameters, which are lag time, moon age, arc of vision, arc of light, difference in azimuth, and contrast. Each of the parameters is evaluated with 8,025 data points of lunar crescent observation, and the result is portrayed in a box and whisker graph to demonstrate the minimum, maximum, mean, first, and third quartiles of the result.

4.4.1 Lag Time

Figure 4.1 demonstrate the lag time of visible lunar crescent observation is based on two methods of observations: Optical-aided observation and naked-eye observation. For optical-aided observation, the mean value is 45.60 minutes with, minimum and maximum values of 2.83 minutes and 109.70 minutes, respectively. The first quartile of the optical-aided observation is at 38.28 minutes, the median at 44.24 minutes, while the third quarter is at 52.16 minutes. On the other hand, naked-eye observation has a mean value of 69.33 minutes, with a minimum and maximum values of 28.63 minutes and 197.46 minutes, respectively. The first quartile of the naked-eye observation is at 56.02, median at 66.72 minutes, while third quarter is at 80.46 minutes.

4.4.2 Moon Age

Figure 4.2 demonstrate that the moon age of visible lunar crescent sightings based on two methods of observation: Optical-aided observation and naked-eye observation. Optical-aided observation has a mean value of 22.35 hours with a minimum and maximum of 5.87 hours and 44.46 hours, respectively. The first quartile of the optical-aided observation is at 17.48 hours, the median is at 21.18 hours, and third quarter is at 25.88 hours. On the other hand, naked-eye observation has a mean value of 30.55 hours, with a minimum and

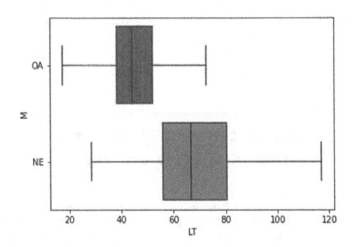

FIGURE 4.1   Box Plot for Lag Time Based on Method of Observation. Source: Researcher Data.

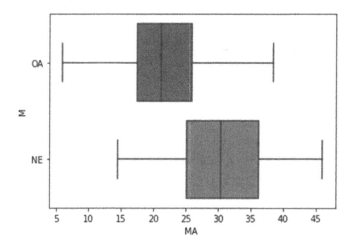

FIGURE 4.2   Box Plot for Moon Age Based on Method of Observation. Source: Researcher Data.

maximum of 14.49 hours and 45.95 hours, respectively. The first quartile of the naked-eye observation is at 25.10 hours, the median at 30.40 hours, and third quarter is at 36.18 hours.

4.4.3 Arc of Vision

Figure 4.3 demonstrate that the arc of vision of visible lunar crescent observation is based on two methods of observation: Optical-aided observation and naked-eye observation.

Optical-aided observation has a mean value of 8.78 degrees, with a minimum and maximum of 0.45 degrees and 20.62 degrees, respectively. The first quartile of the optical-aided observation is at 7.38 degrees, the median is at 8.68 degrees, and the third quarter is at 9.95 degrees. On the other hand, naked-eye observation has a mean value of 13.21 degrees, with a minimum and maximum of 5.29 degrees and 36.78 degrees, respectively. The first quartile of the naked-eye observation is at 10.58 degrees, the median is at 12.59 degrees, and the third quarter is at 15.57 degrees.

4.4.4 Arc of Light

Figure 4.4 demonstrate that the arc of light of visible lunar crescent observation is based on two methods of observation: Optical-aided observation

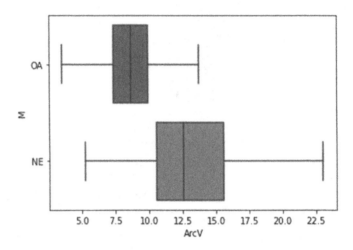

FIGURE 4.3   Box Plot for Arc of Vision Based on Method of Observation. Source: Researcher Data.

and naked-eye observation. Optical-aided observation has a mean value of 11.73 degrees, with minimum and maximum of 4.63 degrees and 30.91 degrees, respectively. The first quartile of the optical-aided observation is at 9.29 degrees, the median is at 10.96 degrees, and the third quarter is at 13.50 degrees. On the other hand, naked-eye observation has a mean value of 16.83 degrees, with a minimum and maximum of 7.39 degrees and 45.14 degrees, respectively. The first quartile of the naked-eye observation is at 13.23 degrees, the median is at 16.26 degrees, and the third quarter is at 19.77 degree.

4.4.5 Difference in Azimuth

Figure 4.5 demonstrate that the difference in azimuth of visible lunar crescent observation is based on two methods of observation: Optical-aided observation and naked-eye observation. Optical-aided observation has a mean value of 6.47 degrees, with minimum and maximum of 0.01 degree and 27.10 degrees, respectively. The first quartile of the optical-aided observation is at 2.44 degrees, the median is at 5.92 degrees, and third quartile is at 9.62 degrees. On the other hand, naked-eye observation has a mean value of 9.04 degrees, with a minimum and maximum of 0.01 degree and 42.75 degrees, respectively. The first quartile of the naked-eye observation is at 4.34 degrees, the median sits at 8.45 degrees, and third quartile is at 12.77 degree.

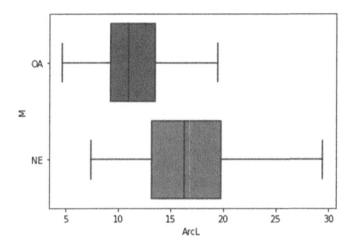

FIGURE 4.4   Box Plot for Arc of Light Based on Method of Observation. Source: Researcher Data.

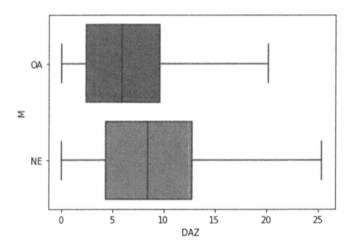

FIGURE 4.5   Box Plot for Difference in Azimuth Based on Method of Observation. Source: Researcher Data.

4.4.6 Width

Figure 4.6 demonstrate that the width of the visible lunar crescent observation is based on two methods of observation: Optical-aided observation and naked-eye observation. Optical-aided observation has a mean value of 21.74 minutes with minimum and maximum of 2.88 minutes and 127.57 minutes, respectively. The first quartile of the optical-aided observation is at 12.30 minutes, while third quarter is at 26.29 minutes. On the other hand, naked-eye observation has a mean value of 43.78 minutes, with a minimum and maximum of 8.74 minutes and 262.17 minutes, respectively. The first quartile of the naked-eye observation is at 25.16 minutes, while the third quarter is at 55.67 minutes. A summary of the boxplot analysis of the lunar crescent visibility criterion can be found in Table 4.7.

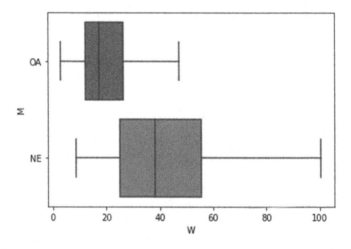

FIGURE 4.6 Box Plot Graph for Width Based on Method of Observation. Source : Researcher Data.

TABLE 4.7 Descriptive Statistics Summary of Lunar Crescent Visibility Variable

| PARAMETER | MEAN | | MIN | | FIRST Q | | THIRD Q | |
	NE	OA	NE	OA	NE	OA	NE	OA
LG (min)	60.44	45.60	28.63	2.83	56.02	38.28	80.46	52.16
MA (hours)	30.55	22.35	14.49	5.87	25.10	17.49	36.18	25.88
ArcV (°)	13.22	8.78	5.29	0.45	10.58	12.59	15.57	9.95
ARCL(°)	16.84	11.73	7.39	4.69	13.23	9.29	19.77	13.50
DAZ (°)	9.05	6.47	0.01	0.01	4.34	2.44	12.77	9.62
Width (°)	43.77	21.74	8.74	2.88	25.16	12.30	55.67	26.30

Source: Researcher Data.

4.5 ANALYSIS OF THE LUNAR CRESCENT VISIBILITY CRITERION[6]

4.5.1 Arc of Vision Over Elongation Lunar Crescent Visibility Criterion

Figure 4.7 demonstrates position of lunar crescent visibility criteria of Istanbul 2015, MABIMS 1995, and MABIMS 2021. Istanbul 2016, MABIMS 1995, and MABIMS 2022 interestingly share the same nature of their contradiction rate. This criterion has a low negative contradiction rate, where Istanbul has 15.61 per cent, MABIMS 1995 has 6.85 and MABIMS 2021 has 6.82 per cent. These criteria are designed to reduce any positive lunar crescent sightings below their visibility line. This is common for lunar crescent visibility criteria adapted for calendrical purposes to avoid errors in positive sightings. MABIMS lunar crescent visibility criterion, which adapted by Brunei, Indonesia, Malaysia, and Singapore for their Islamic calendar regulation, cannot have any single

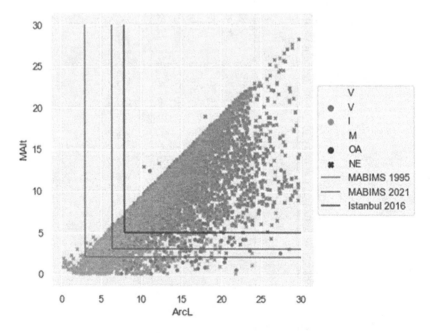

FIGURE 4.7 Scatter Plot for Moon Altitude over Arc of Light Lunar Crescent Visibility Criterion.

TABLE 4.8   Arc of Vision versus Arc of Light Lunar Crescent Visibility Criteria

	ISTANBUL 2016		MABIMS 1995		MABIMS 2021	
	POSITIVE	NEGATIVE	POSITIVE	NEGATIVE	POSITIVE	NEGATIVE
Whole	29.04	15.61	37.29	6.85	33.49	6.82
Naked eye	34.2	2.74	43.27	0.0	39.22	0.18
Optical aided	0.94	100.0	0.90	100.0	0.91	100.0

positive lunar crescent sighting that falls below its visibility line. A single positive lunar crescent observation that falls below the criterion visibility will lead to confusion in determining the exact date of the new Hijri month. Istanbul 2016 functions similarly to the MABIMS lunar crescent visibility criterion to avoid sighted lunar crescent below its visibility line. The contradiction rate of the criteria is portrayed in Table 4.8.

Another trait of criteria designed for calendrical purposes is that it favours naked-eye observation rather than optical-aided observation. This is evidenced in MABIMS 1995 and MABIMS 2021, which have smaller negative naked-eye sighting contradiction rates, while Istanbul has only a 2.74 per cent positive naked-eye sighting contradiction rate. The reason for the criteria favouring naked-eye lunar crescent observation is because optical-aided observation is usually conducted by experts and the technique is not accessible to the public. Naked-eye observation, on the other hand, is capable of being conducted by the masses and is accessible to all layer of the population. A positive naked-eye observation that falls below the visibility limit would gather attention and, subsequently, more confusion among the public. However, a positive optical-aided observation that falls below the visibility limit usually attracts interest among the experts and gather miniature attraction in comparison to naked-eye observation.

Istanbul 2016 lunar crescent visibility criterion is not effective as a lunar crescent visibility criterion, as there are numbers of naked-eye lunar crescent sightings that sits below the criterion visibility line. A naked-eye lunar crescent sighting that goes against a lunar crescent visibility criterion would create confusion in the public. MABIMS 1995, while effective at negating any contradicting positive naked-eye and optical-aided observations, is too close to the horizon and does not reflect an actual observation of the lunar crescent. A Hijri calendar-purposed lunar crescent visibility criterion must reflect an actual observation of the lunar crescent during the twenty-ninth of an Islamic month. MABIMS 2021 is the most suitable criterion for Hijri calendrical purposes. The first reason is that the MABIMS 2021 lunar crescent visibility criterion reflects an actual observation of the lunar crescent, by having a parameter based on the lower limit of naked-eye observation of the lunar crescent. Second, the MABIMS 2021 does not have any positive naked-eye lunar crescent sightings located below its criterion line.

However, there are a number of recommendations to improve MABIMS 2021 as a Hijri calendrical reference. First, MABIMS 2021's 6.4-degree elongation parameter is closely located above the minimum telescopic records of optical-aided observation, which is 5.96 degrees. Currently, there are two telescopic lunar crescent observations that are located below the 6.4-degree elongation limit. Paired with high value of arc of vision and lunar crescent width, a skilful observer is able to break the 6.4-degree elongation parameter limit. As telescopic observation is now becoming norm sighting methodology among religious officials, observatories, and amateur astronomers, a lunar crescent visibility criterion should have an elongation parameter close to the minimum telescopic observation records, in accordance with the lunar crescent observation norm. Should MABIMS 2021 adopt telescopic observation as the criterion reference, a value of 5.50 degrees of elongation is recommended, as it is located between minimum visibility telescopic observation and CCD observation, which are 5.96 degrees and 4.63 degrees, respectively.

Second, the MABIMS 2021 lunar crescent visibility criterion parameter design is not dynamic and unable to follow the changing nature of lunar crescent cycle. The way the MABIMS 2021 is designed is that it follows two logic conditions: Arc of vision condition and the arc of light condition. This kind of design has its flaws, as there will be cases where a sighted lunar crescent has an arc of vision condition that passes MABIMS 2021 but does not pass MABIMS 2021's arc of elongation condition. Alternatively, there will be cases where a sighted lunar crescent has an arc of light higher than the MABIMS 2021 parameter but does not pass MABIMS 2021's arc of vision parameter. These cases are particularly true for locations with high latitude, as elongation value is higher than locations located near the equator. In our data, there are 65 positive lunar crescent observations that do not pass one of MABIMS 2021 criteria, and 3 of the observations are located in Southeast Asia.

To negate this issue, a calendrical lunar crescent visibility criterion should utilize an expression as its design parameter. Expression parameters are more dynamic and able to composite both the arc of vision and the arc of light into a single condition. This reduces the chances of observing a lunar crescent sighted below the criterion line, particularly in high latitude locations. As a suggestion, moon altitude versus arc of light lunar crescent visibility criterion that prioritizes the removal of negative sighting contradictions is expressed as:

$$MAlt = -0.3351\ ArcL + 0.0023\ ArcL2 + 0.000064\ ArcL3 + 7.78 \qquad 4.1$$

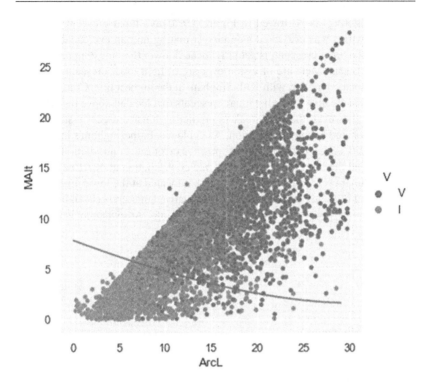

FIGURE 4.8   Scatter Plot Suggested Lunar Crescent Visibility Criterion.

The suggested lunar crescent visibility criterion is able to predict the visibility of the lunar crescent with a 30.414 per cent positive contradiction rate and a 0.0 per cent negative contradiction rate for naked-eye prediction. This is better than MABIMS 2021, MABIMS 1995, and Istanbul 2021 in predicting lunar crescent visibility and is applicable for a wide range of location latitudes. The position of the lunar crescent visibility criterion over data of lunar crescent sightings is portrayed in Figure 4.8.

4.5.2 Arc of Vision over Azimuth Lunar Crescent Visibility Criterion

Arc of vision versus azimuth lunar crescent visibility criterion was introduced by Fotheringham in 1910. Maunder, in 1911, published an improvement over Fotheringham. In the following year, this was followed by Ilyas in 1988, Fatoohi in 1998, and Krauss in 2012. The position of their criteria over data of lunar crescent sightings is portrayed in Figure 4.9. Ilyas's lunar crescent visibility criterion, which was criticized as underestimating human eye capabilities in detecting the lunar crescent, is actually located lower than most lunar crescent visibility criteria. Most arc of vision over arc of light lunar crescent visibility criteria is located higher, with Fotheringham at the highest line. Consequently, these criteria are able to predict lunar crescents that located above the criterion accurately; however, they are weak in predicting lunar crescent visibility criteria that located below the criterion. As evidence, Fotheringham's lunar crescent visibility criterion has the lowest positive error rate contradiction, at 11.46 per cent, and the largest negative error rate contradiction at 46.01 per cent. In contrast, lunar crescent visibility criteria that located at the lower line, such as Fatoohi and Ilyas, are better at predicting negative lunar crescent sightings and worse at predicting positive lunar crescent sightings. As evidence, Fatoohi has

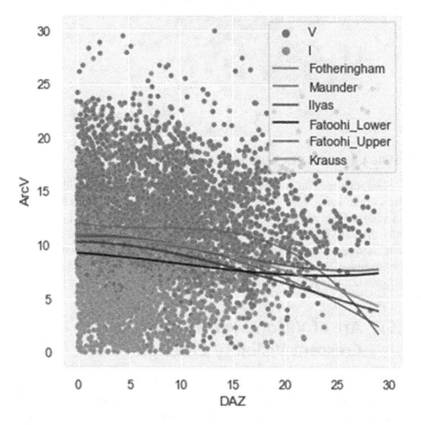

FIGURE 4.9   Scatter Plot for Arc of Vision over Difference in Azimuth Lunar Crescent Visibility Criterion.

25.67 per cent negative error rate contradiction and a 17.28 per cent at positive contradiction rate. The full report of the criteria contradiction rate is portrayed in Table 4.9.

Figure 4.10 demonstrates the position of the criteria over minimum data of lunar crescent sightings. All the arc of vision versus the difference in azimuth lunar crescent visibility criteria ignore optical-aided observation of the lunar crescent sightings. This is because the data of lunar crescent sightings in the database of Fotheringham, Maunder, Ilyas, Fatoohi, and Kraus do not include optical-aided lunar crescent sighting, particularly telescopic, and CCD sightings. The use of telescope and CCD in lunar crescent observation only started primarily in the early twenty-first century. Due to the exclusion of optical-aided lunar crescent sighting, most of the arc of vision versus the difference in azimuth criteria have high value of mean absolute error (MAE) and mean squared error (MSE). This is demonstrated in Table 4.10.

As a suggestion, the arc of vision versus the difference in azimuth lunar crescent visibility criterion that prioritizes the removal of naked-eye negative sighting contradictions is expressed in Equation 4.2.

$$ArcV = -0.09222 \, DAZ + -0.00629 \, DAZ2 + 0.0002078 \, DAZ3 + 6.792 \quad 4.2$$

The suggested lunar crescent visibility criterion is able to predict the visibility of the lunar crescent with a 29.85 per cent positive contradiction rate and a 0.43 per cent negative contradiction rate for naked-eye prediction. This is better at predicting naked-eye lunar crescent visibility and is applicable for a wide range of location latitudes. The position of the lunar crescent visibility criterion over data of lunar crescent sightings is portrayed in Figure 4.11.

For optical-aided sighting, criterion that prioritizes locating around the optical-aided sighting is expressed as Equation 4.3.

$$ArcV = -1.359O \, DAZ + 0.08171O \, DAZ2 + -0.001533O \, DAZ3 + 7.391 \quad 4.3$$

The lunar crescent visibility criterion is better at predicting optical-aided lunar crescent sightings, and it is located around the data of optical-aided sightings. The criterion MAE is 3.24892, and the criterion is 3.24892 and MSE is 26.549, values better than most arc of vision over difference in azimuth lunar crescent visibility criteria. The shape of the criterion is portrayed in Figure 4.12.

TABLE 4.9   Arc of Vision versus Difference in Azimuth Lunar Crescent Visibility Criteria

	FOTHERINGHAM		MAUNDER		ILYAS		FATOOHI$_{LL}$		FATOOHI$_{UL}$		KRAUSS	
	(+)	(−)	(+)	(−)	(+)	(−)	(+)	(−)	(+)	(−)	(+)	(−)
Whole	11.46	46.01	13.09	38.16	15.09	33.31	17.28	25.67	13.24	37.93	13.43	37.13
Naked eye	12.03	33.11	14.05	22.17	16.37	15.71	19.29	8.58	14.22	21.82	14.46	20.76
Optical aided	0.0	98.96	0.0	98.81	1.08	99.19	1.4	99.63	0.0	98.8	0.0	98.78

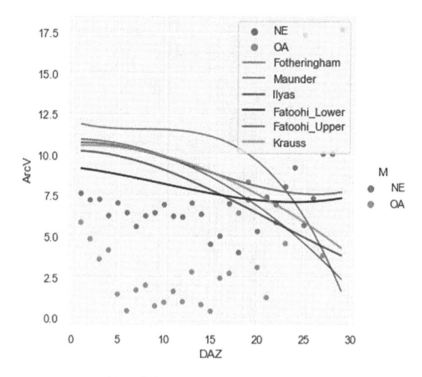

FIGURE 4.10 Position of the Criteria over Minimum Data of Lunar Crescent Sighting.

4.5.3 Arc of Vision over Width Lunar Crescent Visibility Criterion

Arc of vision versus width lunar crescent visibility criterion was introduced by Bruin in 1977. He was the first to introduce width in modern lunar crescent visibility criteria. Bruin's work is then followed by Yallop in 1998, Odeh in 2004, and Qureshi in 2010. The position of their criteria over data of lunar crescent sightings is portrayed in Figure 4.13.

Odeh follows Yallop's model of lunar crescent sighting; therefore, both has the same line with different y-axis starting points. Qureshi and Alrefay add some additional curves in their criterion line; however, it is still has a similar shape to Yallop's and Odeh criteria. In predicting naked-eye sighting, Odeh is the best naked-eye lunar crescent visibility criterion, with a mean percentage of 12.59. Qureshi, however, is worse at predicting naked-eye sighting, with an average percentage of 15.26. Most of the lunar crescent visibility criteria

TABLE 4.10 Regression Analysis for Arc of Vision versus Difference in Azimuth Lunar Crescent Visibility Criteria

CRITERION	FOTHERINGHAM	MAUNDER	ILYAS	FATOOHI$_{LL}$	FATOOHI$_{UL}$	KRAUSS
Mean absolute error	6.13	4.83	4.23	3.55	4.62	4.71
Mean squared error	48.5	33.95	27.11	19.57	29.38	31.6
R-squared	−3.02	−1.81	−1.25	−0.62	−1.44	−1.62

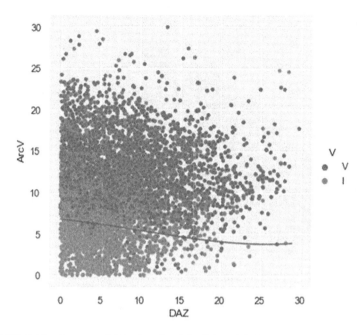

FIGURE 4.11   Suggested Criterion for Naked-Eye Sighting.

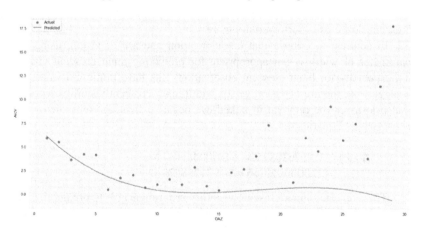

FIGURE 4.12   Suggested Criterion for Optical-Aided Sighting.

are located above the optical-aided lunar crescent sighting, except for Odeh's optical-aided lunar crescent visibility criterion. However, there are a number of lunar crescent sightings observed at lower parameters of arc of vision and

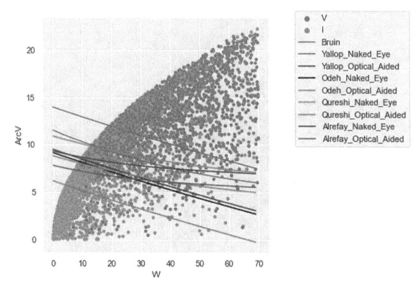

FIGURE 4.13   Scatter Plot for Arc of Vision over Width Lunar Crescent Visibility Criterion.

width, which most of the criteria are unable to predict lunar crescent visibility at lower parameters. The criterion position over the minimum value of lunar crescent sightings is portrayed in Figure 4.14.

In contrast to criteria that use elongation and arc of vision, using the parameter of width is not appropriate for predicting lunar crescent visibility, especially for lunar crescent observations that have minimum value. As a suggestion, the arc of vision versus width lunar crescent visibility criterion that prioritizes the removal of naked-eye negative sighting contradictions is expressed as Equation 4.4.

$$ArcV = -0.0676208958w + 0.0004860914 \ w^2$$
$$+ 0.0000059166 \ w^3 + 8.1128857676 \qquad 4.4$$

The suggested lunar crescent visibility criterion is able to predict the visibility of the lunar crescent with a 33.92 per cent positive contradiction rate and a 4.15 per cent negative contradiction rate for naked-eye prediction. This is better at predicting naked-eye lunar crescent visibility and is applicable for a wide range of location latitudes, with only 84 positive naked-eye lunar crescent sightings located below the criterion. The position of the lunar crescent visibility criterion over data of lunar crescent sightings is portrayed in Figure 4.15.

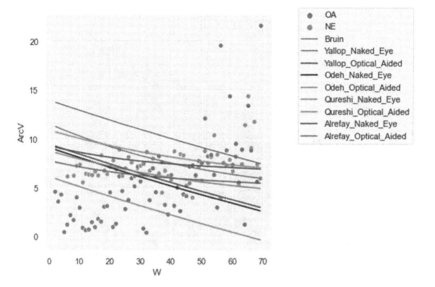

FIGURE 4.14   Scatter Plot of Minimum Data for Arc of Vision over Width Lunar Crescent Visibility Criterion.

For optical-aided sighting, criterion that prioritizes locating around the optical-aided sighting is expressed as Equation 4.5:

$$ArcV = -0.2372w + 0.0064324w2 + -0.0000523w2 + 2.957 \qquad 4.5$$

The lunar crescent visibility criterion is better at predicting optical-aided lunar crescent sightings, and it is located around the data of optical-aided sightings. The criterion MAE is 3.24892, the criterion is 3.73, and MSE is 22.49, values better than most arc of vision over width lunar crescent visibility criterion for optical-aided observation. The shape of the criterion is portrayed in Figure 4.16.

4.5.4 Lag Time Lunar Crescent Visibility Criterion

Generally, lag time is considered a weak parameter for the lunar crescent sighting. Schaefer (1996) highlighted that lag time has consistently high error throughout the parameter and notes that lag time is less predictable at high latitudes. Ahmad et al. (2020) and Ilyas (1983) argue that lag time is only suitable for explaining the visibility of the lunar crescent to laymen and not suitable as

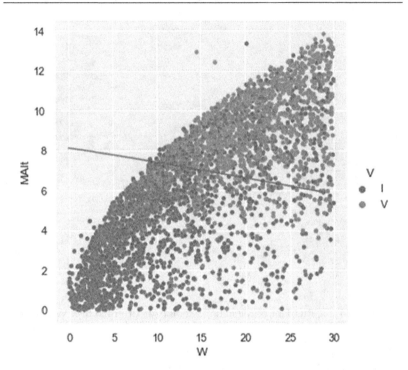

FIGURE 4.15   Recommended Criterion for Naked-Eye Sighting.

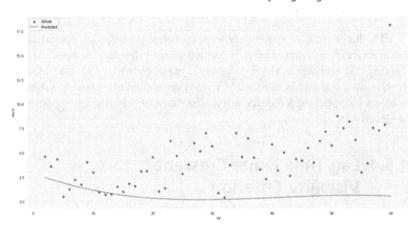

FIGURE 4.16   Recommended Criterion over Minimum Optical-Aided Sighting.

LT						
MA	0.93					
ArcV	0.22	0.2				
ArcL	0.2	0.28	0.72			
DAZ	0.11	0.23	0.074	0.7		
W	0.18	0.26	0.55	0.9	0.74	
V	0.19	0.18	0.53	0.5	0.2	0.35
	LT	MA	ArcV	ArcL	DAZ	W	V

FIGURE 4.17 Pearson Correlation Result of the Lunar Crescent Visibility Parameter.

a lunar crescent visibility criterion. Figure 4.17 demonstrates the correlation result of the lunar crescent visibility parameter. LT, MA, ArcV, ArcL, DAZ, W, and V are represented by lag time, moon age, arc of vision, arc of light, difference in azimuth, width, and visibility, respectively.

On visibility correlation, lag time and difference in azimuth have the weakest correlation, with value of 0.19 and 0.20, respectively. This shows that lag time and difference in azimuth have the weakest correlation influence on lunar crescent visibility in comparison to others. This shows that they do not play as a primary factor in determining the visibility of a lunar crescent. Despite criticism of the lag time parameter in lunar crescent visibility, there are scholars who adopted lag time in their criteria. Caldwell adopted lag time for his lunar crescent visibility criterion. He argued that lag time is more suitable for various latitudes of observation sites. This is because while the arc of vision is relatively the same at various latitudes, lag time increases significantly at higher latitudes as the path of the lunar crescent is more slanted to the horizon. This increases the chance of visibility. The arc of vision unable to reproduce the slanted path of the lunar crescent; therefore, it is not suitable for lunar crescent visibility criterion.

Figure 4.18 demonstrates weakness of Caldwell's lunar crescent visibility criterion. Caldwell's lunar crescent visibility is limited to data of a lunar crescent below 72 minutes for naked-eye observation and 20 lag time for optical-aided observation. This does not represent the actual observation of the lunar crescent, where lag time ranges from 2 minutes to over 140 minutes. The primary reason for Caldwell's weakness is the incompatibility of the ranges between arc of light and lag time as a parameter for the lunar crescent visibility criterion. The arc of light has a minimum positive observation of

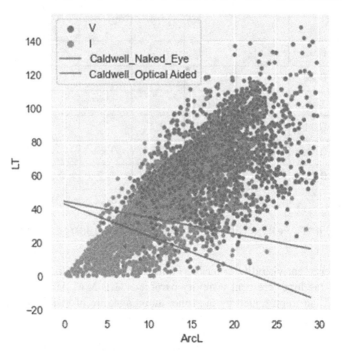

FIGURE 4.18   Caldwell Lunar Crescent Visibility Criterion.

4.63 degrees and a maximum of 45.14 degrees, while lag time has a minimum of 2.83 minutes and a maximum of 197.46 minutes. This huge difference in ranges causes Caldwell's lunar crescent visibility criterion to be limited to data of a lunar crescent below 72 minutes for naked-eye observation and 20 lag time for optical-aided observation. Another reason is that the arc of light distribution is similar to the lag time distribution. This is because lag time is directly proportional to the arc of light; lunar crescents that have higher lag time usually have higher arcs of light, and lunar crescents have lower lag time usually have lower arc of light.

On the other hand, Gautschy was able to produce a more robust lunar crescent visibility criterion using a combination of lag time and difference in azimuth. While differences in azimuth have an incompatible range with lag time, most of the Gautschy data of lunar crescent sightings have a lower value of lag time, with a maximum of 40 minutes. This negates the mismatch range issue between differences in azimuth and lag time. In addition, lag time is not directly proportional to differences in azimuth. A higher lag time does not necessarily translate into a higher value of difference in azimuth. As a

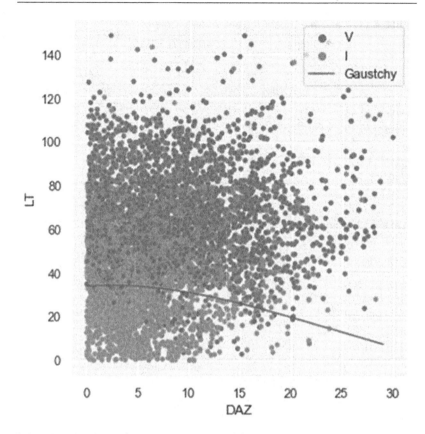

FIGURE 4.19 Gautschy Lunar Crescent Visibility Criterion.

result, Gautschy's lunar crescent visibility criterion line position is better than Caldwell's lunar crescent visibility criterion, as portrayed in Figure 4.19.

Table 4.11 demonstrated that Caldwell's lunar crescent visibility criterion is weaker at predicting lunar crescent visibility in comparison to the Gautschy criterion. The Caldwell criterion line is located far from the data of lunar crescent sightings, causing high mean square error. In comparison, the Gautschy lunar crescent visibility criterion is located around the data of lunar crescent sightings and consequently has a lower mean square error, as illustrated in Table 4.12.

TABLE 4.11   Lag Time Lunar Crescent Visibility Criteria

	CALDWELL NAKED EYE		CALDWELL OPTICAL AIDED		GAUTSCHY	
	POSITIVE	NEGATIVE	POSITIVE	NEGATIVE	POSITIVE	NEGATIVE
Whole	22.48	11.17	32.12	4.72	24.67	8.82
Naked eye	26.39	1.55	37.68	0.0	29.05	0.82
Optical aided	1.08	100.0	0.94	100.0	1.02	100.0

TABLE 4.12   Regression Analysis of Lag Time Criteria

CRITERION	CALDWELL NAKED EYE	CALDWELL OPTICAL AIDED	GAUTSCHY
Mean absolute error (MAE)	38.98	99.36	16.66
Mean squared error (MSE)	2,495.23	14,415.61	578.76
R-squared	−80.9	−472.14	−0.73

4.5.5 Negative Naked-Eye Lunar Crescent Sighting Outliers

In theory, lunar crescents that have high visibility parameters should be able to be sighted by the naked eye. However, there are cases where the lunar crescent is not sighted, despite being situated high above the visibility line. These cases are called negative naked-eye lunar crescent observation outliers. Defining the threshold for outliers might vary from one statistical dataset to another. Choosing outliers based on a specific criterion will bias the result in accordance with the specific criterion, making the definition of outliers vary from one to another. In these cases, the outlier is selected based on two data points: The lunar crescent observation correlation parameter and its quartile location. This ensures the selection is free from bias and based on the actual scatter of the data.

Arc of vision and the arc of light are selected as they have the highest correlation value in determining lunar crescent visibility based on Figure 4.17. Their limiting values for the upper second quartile are 16 degrees and 12 degrees for arc of vision and the arc of light, respectively. The assumption is that the lunar crescent is expected to be sighted at lunar crescents above their upper second quartile parameters. The result of the filtering is 1,268 data points of lunar crescent observations, with 58 data points of negative lunar crescent observations. The scatter plot of the negative lunar crescent observation is portrayed in Figure 4.20.

Addition tests are conducted to further understand its outlier nature. The first test is the comparison with the width parameter. If the lunar crescent has a parameter of width below the upper first quartile of the parameter based on Figure 4.6 and the lunar crescent is classified as 'L', signifying the lower limit of the first quartile. The second test is to check the dating accuracy of the data. This involves determining if the date of the data is incorrectly reported, which could occur due to data date validity issues (common in the case of ancient data), time zone discrepancies, or inconsistencies between calculated Julian dates and reported Julian data. This second test will be classified as 'ID', signifying incorrect data. The third test involves weather and atmospheric condition reports. The International Lunar Crescent Project report provides information about the weather and atmospheric conditions during the observation. The third test will be classified as 'C', signifying cloudy. Data that does

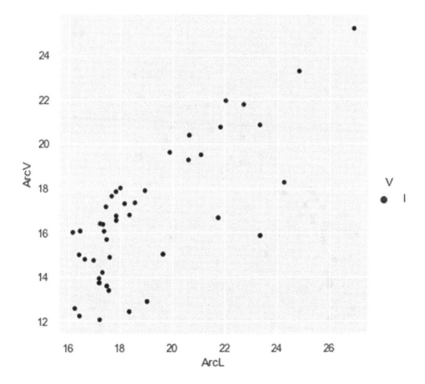

FIGURE 4.20   Negative Lunar Crescent Sighting Outlier.

not fall under the conditions of these three tests is classified as 'U', signifying upper limit. The result of the additional tests is as shown in Figure 4.21.

There are nine data classified as incorrect dating, from Krauss and Fatoohi's Babylonian data. Krauss conceded that Babylonian recorded lunar crescent observation has a 1-day contradiction rate, making it susceptible to being classified as outliers. A total of 21 data points located at the lower limit from the upper first quarter value of width and difference in azimuth. Width is directly proportional to the surface area of the moon, while differences in azimuth correspond to the brightness of the background twilight sky. Thus, these 21 data points are classified as outliers as they have low contrast value to be detected by the human naked eye. A total of 23 points were observed during cloudy weather and hazy atmospheric conditions, making the lunar crescent not visible during observation. There are five data points not sighted, despite being located above all parameter visibility lines and having cloudless sky conditions.

Schaefer has highlighted that eyesight, experience, and age contribute to the probability of successful sighting. Although the effect is small, it is

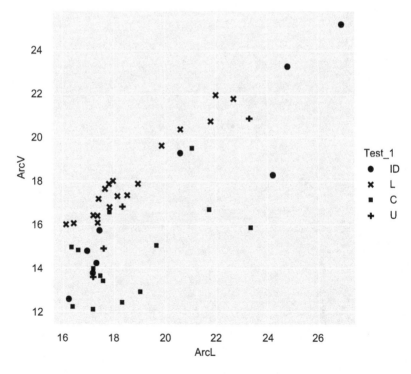

FIGURE 4.21   Negative Lunar Crescent Sighting Outliers with Addition Test.

non-negligible and present a real impact on sighting probability. Schaefer also added that errors in locating the moon position contribute to at least 2 per cent of the total collected data (Doggett et al.(1994). In this case, all of the observers are members of ICOP and are well-known experts in lunar crescent visibility among their community. However, as mentioned by Schaefer, this observation could fall into the categorization of 2 per cent human error from the total collected data.

4.6 CONCLUSION

A total of 8,290 records of lunar crescent observation data were collected for the purpose of analysis. The analysis tool, named HilalPy, is able to demonstrate positive and negative contradiction rates of a lunar crescent visibility criterion, using 8,001 records of lunar crescent observation data. HilalPy enables the endeavour of a standard comparative analysis of one lunar crescent visibility criterion to another, as the data of lunar crescent sightings and its computed values have been designated as constant variables. For example, the newly introduced MABIMS 2021 lunar crescent visibility criterion can be compared with the internationally agreed Istanbul Declaration 2016 lunar crescent visibility criterion. HilalPy's main purpose is to be adopted as a standard for measuring the reliability of a new lunar crescent visibility criterion.

NOTES

1. M. S. Faid, M. S. A. Mohd Nawawi, and M. H. Mohd Saadon, "Analysis Tool for Lunar Crescent Visibility Criterion based on Integrated Lunar Crescent Database," *Astronomy and Computing* 45 (2023): 11.
2. Susiknan Azhari, "Gagasan Menyatukan Umat Islam Indonesia Melalui Kalender Islam [The Idea of Uniting Indonesian Muslims Through the Islamic Calendar]," *AHKAM: Jurnal Ilmu Syariah* (2015), https://doi.org/10.15408/ajis.v15i2.2869; Susiknan Azhari, *Kalendar Islam: Ke Arah Integrasi Muhammadiyah-NU [Islamic Calendar: Towards Muhammadiyah-NU Integration]* (Yogjakarta: Museum Astronomi Islam, 2012).
3. Muhamad Syazwan Faid et al., "Confirmation Methodology for a Lunar Crescent Sighting Report," *New Astronomy* 103 (2023): 4.

4. T. Alrefay et al., "Analysis of Observations of Earliest Visibility of the Lunar Crescent," *The Observatory* 138, no. 1267 (2018); Mohammad Odeh, "New Criterion for Lunar Crescent Visibility," *Experimental Astronomy* 18 (2004): 39–64; Louay Fatoohi, "First Visibility of the Lunar Crescent and Other Problems in Historical Astronomy" (PhD, University of Durham, 1998); Andrew Crumey, "Human Contrast Threshold and Astronomical Visibility," *Monthly Notices of the Royal Astronomical Society* 442, no. 3 (2014), https://doi.org/10.1093/mnras/stu992; Bradley Schaefer, "Lunar Crescent Visibility," *Quarterly Journal of Royal Astronomical Society* (1996): 759–68; M. Ilyas, "Lunar Crescent Visibility Criterion and Islamic Calendar," *Quarterly Journal of Royal Astronomical Society* 35 (1994): 425–61.
5. Muhamad Syazwan Faid et al., "Assessment and Review of Modern Lunar Crescent Visibility Criterion," *Icarus* 412 (2024); M. S. Faid, M. S. A. Mohd Nawawi, and M. H. Mohd Saadon, "Analysis Tool for Lunar Crescent Visibility Criterion based on Integrated Lunar Crescent Database," *Astronomy and Computing* 45 (2023): 100752–52.
6. Muhamad Syazwan Faid et al., "Assessment and Review of Modern Lunar Crescent Visibility Criterion," *Icarus* 412 (2024); M. S. Faid, M. S. A. Mohd Nawawi, and M. H. Mohd Saadon, "Analysis Tool for Lunar Crescent Visibility Criterion based on Integrated Lunar Crescent Database," *Astronomy and Computing* 45 (2023): 100752–52.

Conclusion and Summary

5

5.1 REVIEW ON THE LUNAR CRESCENT VISIBILITY CRITERION

1. Lunar crescent visibility parameter's strengths and weaknesses are demonstrated. While altitude, lag time, and moon age play a simpler role in determining lunar crescent visibility, it is not entirely dependable as altitude, lag time, and moon age parameter are prone to error at higher geographical latitudes. Literature suggests that elongation is the most reliable lunar crescent visibility criterion; however, elongation-limiting visibility is heterogeneous among researchers of lunar crescent visibility, ranging from 5 degree of elongation to 10.5 degree of elongation.

2. Modern lunar crescent visibility criterion literature review demonstrated that the two-parameter lunar crescent visibility criterion is the most popular adaptation for lunar crescent visibility criterion research. This includes Fotheringham in 1910 and Alrefay in 2018. Odeh introduced a multi-range lunar crescent visibility criterion in 1998. While the first iteration of multi-range lunar crescent visibility criterion has blurry lines between one visibility to another, other researchers have simplified the multi-range lunar crescent visibility criterion into naked-eye visibility and optical-aided visibility, as shown by Alrefay and Ahmad. Fatoohi used multi-range lunar crescent visibility criterion to demonstrate the concept of uncertainty zone, a concept first highlighted by Ilyas in 1993.

3. Contradiction rate has proven to be the most appropriate assessment for lunar crescent visibility criterion, followed by literature analysis because contradiction rate is performed based on actual data of lunar crescent records. Lunar cycle analysis does not have its impracticality impact, while histogram bias analysis only caters for

DOI: 10.1201/9781003536192-5

single parameter type criterion. The discussion also highlights that while the most comprehensive assessment on lunar crescent visibility criterion is Fatoohi, the assessment is outdated since it was performed in 1998. Since its publication, there are numerous reports of recorded lunar crescent observations and publications of lunar crescent visibility criteria that need to be assessed so that they can be compared analytically from one to another.

5.2 ASSESSMENT OF PERCENTAGES OF RELIABILITY OF LUNAR CRESCENT VISIBILITY CRITERION

1. Differences in azimuth and moon age are ineffective parameters of lunar crescent sighting. This is because the difference in azimuth has a similar minimum value for both naked-eye observation and optical-aid observation, while the moon age has a large disparity between the minimum value for both naked-eye observation and optical-aid observation. Moon age and difference in azimuth also have low correlation strength towards visibility. Arc of light, followed by arc of vision, is the most suitable parameter for lunar crescent sighting due to the moderate disparity between naked-eye observation and optical-aid observation, followed by a stronger correlation towards lunar crescent visibility.

2. For equation-style lunar crescent visibility criterion, $Fatoohi_{Lower\,Limit}$ is the most suitable for naked-eye lunar crescent visibility criterion, having a contradiction rate of 19.5 percent for positive contradiction rate, and 8.82 percent for negative contradiction rate. $Fatoohi_{Lower\,Limit}$ criterion, published in 1998, outperformed its predecessor, including Odeh and Caldwell. $Fatoohi_{Lower\,Limit}$ performance is due to the integration of an uncertainty zone and the introduction of lower and upper limits of lunar crescent visibility criterion that are located outside zone of uncertainty. Alrefay is the most suitable for optical-aided observation, having a contradiction rate of 14.16 percent for positive contradiction rate and 22.33 percent for negative contradiction rate. The primary reason for this performance is that Alrefay's data includes latest data of telescopic observation. In addition, Alrefay outlines a clear definition between optical-aided and naked-eye observations for lunar crescent visibility criterion, in

contrast to the convoluted and multi-range lunar crescent visibility criterion of Yallop, Odeh, and Qureshi, allowing Alrefay to perform a more accurate optical-aided lunar crescent visibility criterion.

3. MABIMS 2021 is the most suitable criterion for Hijri calendrical purpose because MABIMS 2021 has a zero-contradiction rate for negative naked-eye observation, and its parameter reflects an actual observation of the lunar crescent. On the contrary, Istanbul Declaration 2016 and MABIMS 1995 are not ideal criteria for Hijri calendrical purpose because Istanbul 2016 criterion has a 0.93 percent negative naked-eye contradiction rate or eight records of positive naked-eye lunar crescent sightings that are located below the Istanbul 2016 criterion line. The MABIMS 1995 lunar crescent visibility criterion is too close to the horizon and does not mirror a lunar crescent observation. However, MABIMS 2021 could be improved by lowering the elongation parameter to 5.50 degrees, a value that is located between minimum telescopic observation (5.96 degree) and CCD observation (4.63 degree). In addition, design change from a two-parameter conditional style lunar crescent visibility criterion to a singular equation style will make MABIMS 2021 more dynamic towards changes in lunar crescent parameter and more accurate for Hijri calendrical application.

The assessment of lunar crescent visibility criterion is vital for Hijri calendar determination. Ilyas, Schaefer, and Fatoohi have previously conducted research on this matter to provide insight into the reliability of the criterion. However, most of their research predates 1998 and requires a refreshed view since there are more data and visibility criteria since their published analysis. To reach the endeavour, this chapter aims to provide a comparative analysis of lunar crescent visibility criterion based on 8,290 records, including 5,267 positive lunar crescent sightings and 3,023 negative records. The analysis is conducted using swarm plot analysis, contradiction rate analysis, and regression analysis. Analysis on the arc of vision versus elongation lunar crescent visibility criterion, which is favourably used in most of the Islamic countries, found that it is designed to eliminate confusion in naked-eye observation of the lunar crescent. However, the current criterion is based on a logical expression instead of dynamic expression, which does not work well in high latitude locations. The arc of vision versus azimuth criterion assessment found that it primarily has high success rates in predicting positive lunar crescent sightings for the Fotheringham and Maunder criterion, and works well for predicting negative lunar crescent sightings for Fatoohi and Ilyas. The azimuth-arc of vision criterion found diminishing popularity in recent years; therefore, it does not include critical optical-aided observation. Bruin pioneered the use of width in lunar crescent visibility criteria, and Yallop sparked the interest in multi-range lunar crescent visibility criteria among researchers, such as Odeh and Qureshi. The width lunar crescent visibility criterion is inconsistent from one lunar cycle to another, making it a less reliable criterion for predicting lunar crescent visibility. The lag time lunar crescent visibility criterion faced criticism among researchers in the late twentieth century; however, Gautschy was able to demonstrate the strength of the lag time variable in predicting lunar crescent sighting, particularly for naked-eye observation. This chapter also found that there is at least 2 percent error in locating the visibility of the lunar crescent among expert observers, echoing the finding of Doggett et al. in 1994. The chapter also provides an alternative lunar crescent visibility criterion for each of the criterion models, hoping to ignite an engagement in determining the best model for lunar crescent sighting.

Bibliography

Ahmad, Nazhatulshima, Nur Izzatul Najihah Mohamad, Raihana Abdul Wahab, Mohd Saiful Anwar Mohd Nawawi, Mohd Zambri Zainuddin, and Ibrahim Mohamed. "Analysis Data of the 22 Years of Observations on the Young Crescent Moon at Telok Kemang Observatory in Relation to the Imkanur Rukyah Criteria 1995." *Sains Malaysiana* 51, no. 10 (2022): 3415–22. https://doi.org/10.17576/jsm-2022 -5110-24.

Ahmad, Nazhatulshima, Mohd Saiful Anwar Mohd Nawawi, Mohd Zambri Zainuddin, Zuhaili Mohd Nasir, Rossita Mohamad Yunus, and Ibrahim Mohamed. "A New Crescent Moon Visibility Criteria Using Circular Regression Model: A Case Study of Teluk Kemang, Malaysia." *Sains Malaysiana* 49, no. 4 (2020): 859–70. https://doi.org/10.17576/jsm-2020-4904-15.

Alrefay, T., S. Alsaab, F. Alshehri, A. Alghamdi, A. Hadadi, M. Alotaibi, K. Almutari, and Y. Mubarki. "Analysis of Observations of Earliest Visibility of the Lunar Crescent." *The Observatory* 138, no. 1267 (2018): 267–91.

Azhari, Susiknan. "Cabaran Kalendar Islam Global Di Era Revolusi Industri 4.0." *Jurnal Fiqh* 18, no. 1 (2021): 117–34. https://doi.org/10.22452/fiqh.vol18no1.4. https://fiqh.um.edu.my/index.php/fiqh/article/view/30691.

Azhari, Susiknan. "Gagasan Menyatukan Umat Islam Indonesia Melalui Kalender Islam [The Idea of Uniting Indonesian Muslims through the Islamic Calendar]." *AHKAM: Jurnal Ilmu Syariah* (2015). https://doi.org/10.15408/ajis.v15i2.2869.

Azhari, Susiknan. *Kalendar Islam: Ke Arah Integrasi Muhammadiyah-Nu [Islamic Calendar: Towards Muhammadiyah-Nu Integration]*. Yogjakarta: Museum Astronomi Islam, 2012.

Bemporad, A. "La Teoria Della Estinzione Atmosferica Nella Ipotesi Di Un Decrescimento Uniforme Della Temperatura Dell'aria Coll'altezza." *Memorie della Societa Degli Spettroscopisti Italiani* 33 (1904): 31–37.

"Beautiful Soup Documentation." Crummy, 2020. https://www.crummy.com/software /BeautifulSoup/bs4/doc/.

Blackwell, H. R. "Contrast Thresholds of the Human Eye." *Journal of the Optical Society of America* 36, no. 11 (1946): 624–43. https://doi.org/10.1364/JOSA.40 .000825.

Bruin, Frans. "The First Visibility of the Lunar Crescent." *Vistas in Astronomy* 21 (1977): 331–58. https://doi.org/10.1016/0083-6656(77)90021-6. https://linking-hub.elsevier.com/retrieve/pii/0083665677900216.

Caldwell, John. "Moonset Lag with Arc of Light Predicts Crescent Visibility." *Monthly Notes of the Astronomical Society of South Africa* 70, no. 11&12 (2011): 220–35.

"Computational Astronomy and the Earliest Visibility of Lunar Crescent." International Crescent Observation Project, Updated 2005. http://www.icoproject.org/paper .html.

Crumey, Andrew. "Human Contrast Threshold and Astronomical Visibility." *Monthly Notices of the Royal Astronomical Society* 442, no. 3 (2014): 2600–19. https://doi.org/10.1093/mnras/stu992.

Danjon, Andre-Loius. "Le Croissant Lunaire [the Lunar Crescent]." *l'Astronomie: Bulletin de la Société Astronomique de France* 50 (1936): 57–65.

Doggett, L. E., P. K. Seidelmann, and B. E. Schaefer. "Moonwatch - July 14, 1988." *Sky and Telescope* 76 (1988): 34–35.

Doggett, L., Bradley Schaefer, and L. Doggett. "Lunar Crescent Visibility." *Icarus* 107, no. 2 (1994): 388–403. https://doi.org/10.1006/icar.1994.1031.

Doggett, Leroy, and Bradley Schaefer. "Result of the July Moonwatch." In *Sky & Telescope*, 373–75. Florida: Sky Publishing, 1989.

Eddy, John A. "The Maunder Minimum." *Science* 192, no. 4245 (1976): 1189–202.

"Extracttable-Py: Python Library to Extract Tabular Data from Images and Scanned Pdfs." Github, 2022. https://github.com/ExtractTable/ExtractTable-py.

Faid, M. S., M. S. A. Mohd Nawawi, and M. H. Mohd Saadon. "Analysis Tool for Lunar Crescent Visibility Criterion Based on Integrated Lunar Crescent Database." *Astronomy and Computing* 45 (2023): 100752–52. https://doi.org/10.1016/j.ascom.2023.100752.

Faid, Muhamad Syazwan, Mohd Saiful Anwar Mohd Nawawi, Raihana Abdul Wahab, and Nazhastulshima Ahmad. "Hilalpy: Software to Analyse Lunar Sighting Criteria." *Software Impacts* 18 (2023): 100593–93. https://doi.org/10.1016/j.simpa.2023.100593.

Faid, Muhamad Syazwan, Mohd Saiful Anwar Mohd Nawawi, and Mohd Hafiz Mohd Saadon. "Hilalpy: Analysis Tool for Lunar Crescent Visibility Criterion." ascl:2307.031Astrophysics Source Code Library, 2023.

Faid, Muhamad Syazwan, Mohd Saiful Anwar Mohd Nawawi, Muhammad Syaoqi Nahwandi, Nur Nafhatun Md Shariff, and Zety Sharizat Hamidi. "Hilal-Obs: Authentication Agorithm for New Moon Visibility Report." Astrophysics Source Code Library, 2021. ascl:2104.003.

Faid, Muhamad Syazwan, Mohd Saiful Anwar Mohd Nawawi, Mohd Hafiz Mohd Saadon, Muhammad Syaoqi Nahwandi, Nur Nafhatun Md Shariff, Zety Sharizat Hamidi, Raihana Abdul Wahab, Mohd Paidi Norman, and Nazhatulshima Ahmad. "Confirmation Methodology for a Lunar Crescent Sighting Report." *New Astronomy* 103 (2023): 102063–63. https://doi.org/10.1016/j.newast.2023.102063.

Faid, Muhamad Syazwan, Mohd Saiful Anwar Mohd Nawawi, Mohd Hafiz Mohd Saadon, Raihana Abdul Wahab, Nazhatulshima Ahmad, Muhamad Syaoqi Nahwandi, Ikramullah Ahmed, and Ibrahim Mohamed. "Assessment and Review of Modern Lunar Crescent Visibility Criterion." *Icarus* 412 (2024). https://doi.org/10.1016/j.icarus.2024.115970.

Faid, Muhamad Syazwan, Nur Nafhatun Md Shariff, Zety Sharizat Hamidi, Norihan Kadir, Nazhatulshima Ahmad, and Raihana Abdul Wahab. "Semi Empirical Modelling of Light Polluted Twilight Sky Brightness." *Jurnal Fizik Malaysia* 39, no. 2 (2018): 30059–67.

Fatoohi, Louay. "First Visibility of the Lunar Crescent and Other Problems in Historical Astronomy." PhD, University of Durham, 1998.

Fatoohi, Louay, F. Richard Stephenson, and Shetha S. Al-Dargazelli. "The Danjon Limit of First Visibility of the Lunar Crescent." *The Observatory* 118 (1998): 65–72.

Fotheringham, J. K. "On the Smallest Visible Phase of the Moon." *Monthly Notices of the Royal Astronomical Society* 70 (1910): 527–27.

Fotheringham, J. K. "The Visibility of the Lunar Crescent." *The Observatory* 44 (1921): 308–11.

Gautschy, Rita. "On the Babylonian Sighting-Criterion for the Lunar Crescent and Its Implications for Egyptian Lunar Data." *Journal for the History of Astronomy* 45, no. 1 (2014): 79–90. https://doi.org/10.1177/002182861404500105.

Gautschy, Rita, Michael E. Habicht, Francesco M. Galassi, Daniela Rutica, Frank J. Rühli, and Rainer Hannig. "A New Astronomically Based Chronological Model for the Egyptian Old Kingdom." *Journal of Egyptian History* 10, no. 2 (2017): 69–108. https://doi.org/10.1163/18741665-12340035.

Gautschy, Rita, and Johannes Thomann. "Dating Historical Arabic Observations." *Proceedings of the International Astronomical Union* 14, no. A30 (2020): 163–66. https://doi.org/10.1017/s1743921319003983.

Hasanzadeh, Amir. "Study of Danjon Limit in Moon Crescent Sighting." *Astrophysics and Space Science* 339, no. 2 (2012): 211–21. https://doi.org/10.1007/s10509-012-1004-y. http://www.springerlink.com/index/10.1007/s10509-012-1004-y.

Hassan-Bello, Abdulmajeed Bolade. "Sharia and Moon Sighting and Calculation Examining Moon Sighting Controversy in Nigeria." *Al-Ahkam* 30, no. 2 (2020): 215–52.

Ilyas, M. "Age as a Criterion of Moon's Earliest Visibility." *The Observatory* 103 (1983): 26–29.

Ilyas, M. "The Danjon Limit of Lunar Visibility: A Re-Examination." *Journal of Royal Astronomy Society Canada* 77, no. 4 (1983): 214–18.

Ilyas, M. "Limb Shortening and the Limiting Elongation for the Lunar Crescent's Visiblity." *Quarterly Journal of Royal Astronomy Society* 25 (1984): 421–22.

Ilyas, M. "Lunar Calendars: The Missing Datelines." *The Journal of the Royal Astronomical Society of Canada* 80 (1986): 328–35.

Ilyas, M. "Lunar Crescent Visibility Criterion and Islamic Calendar." *Quarterly Journal of Royal Astronomical Society* 35 (1994): 425–61.

Ilyas, M. *Sistem Kalendar Islam Dari Perspektif Astronomi [The Islamic Calendar System from an Astronomical Perspective]*. Kuala Lumpur: Dewan Bahasa dan Pustaka, 1997.

Ilyas, Mohammad. "Ancients' Criterion of Earliest Visibility of the Lunar Crescent-How Good Is It?" Paper presented at the Proceedings of an International Astronomical Union Colloquium, New Delhi, India, 1987.

Ilyas, Mohammad. *Astronomy of Islamic Times for the Twenty-First Century.* UNKNO, 1988.

Kastner, S. O. "Calculation of the Twilight Visibility Function of Near-Sun Objects." *The Journal of the Royal Astronomical Society of Canada* 70, no. 4 (1976): 153–68.

King, David A. *Astronomy in the Service of Islam.* London: Routledge, 1993.

King, David A. "Frans Bruin (1922–2001)." *Journal for the History of Astronomy* (2002). https://doi.org/10.1177/002182860203300210.

Koomen, M. J., C. Lock, D. M. Packer, R. Scolnik, R. Tousey, and E. O. Hulburt. "Measurements of the Brightness of the Twilight Sky." *Journal of the Optical Society of America* 42, no. 5 (1952): 353–56.

Kordi, Ayman. "The Psychological Effect on Sighting of the New Moon." *The Observatory* 123 (2003): 219–23.

Krauss, Rolf. "Babylonian Crescent Observation and Ptolemaic-Roman Lunar Dates." *PalArch's Journal of Archaeology of Egypt/Egyptology* 9, no. 5 (2012).

Krisciunas, Kevin, and Bradley Schaefer. "A Model of the Brightness of Moonlight." *Publications of the Astronomical Society of the Pacific* 103 (1991): 1033–39.

Loewinger, Y. "Comments on Bradley Schaefer 1988." *Quarterly Journal of Royal Astronomical Society* 36 (1995): 449–52.

Maskufa, Maskufa, Sopa Sopa, Sri Hidayati, and Adi Damanhuri. "Implementation of the New Mabims Crescent Visibility Criteria: Efforts to Unite the Hijriyah Calendar in the Southeast Asian Region." *AHKAM: Jurnal Ilmu Syariah* 22, no. 1 (2022). https://doi.org/10.15408/ajis.v22i1.22275.

Maunder, E. Walter. "On the Smallest Visible Phase of the Moon." *The Journal of the British Astronomical Association* 21 (1911): 355–62.

McNally, D. "The Length of the Lunar Crescent." *Quarterly Journal of Royal Astronomy Society* 24, no. 4 (1983): 417–29.

McPartlan, M. A. "Astronomical Calculation of New Crescent Visibility." *Quarterly Journal of Royal Astronomical Society* 37 (1996): 837–42.

Meeus, Jean. *Astronomical Algorithms.* Virginia: Willmann-Bell, 1991.

Mohd Nawawi, Mohd Saiful Anwar, Saadan Man, Mohd Zambri Zainuddin, Raihana Abdul Wahab, and Nurulhuda Ahmad Zaki. "Sejarah Kriteria Kenampakan Anak Bulan Di Malaysia." *Journal of Al-Tamaddun* 10, no. 2 (2015/12/31): 61–75. https://doi.org/10.22452/JAT.vol10no2.5. https://ejournal.um.edu.my/index.php/JAT/article/view/8690.

Mommsen, August. *Chronologie. Untersuchungen Über Das Kalenderwesen Der Griechen Insonderheit Der Athener.* Leipzig: H.G. Teubner, 1883.

Moosa, Ebrahim. "Shaykh Aḥmad Shākir and the Adoption of a Scientifically-Based Lunar Calendar." *Islamic Law and Society* 5, no. 1 (1998): 57–89.

Mostafa, Zaki. "Lunar Calendars: The New Saudi Arabian Criterion." *The Observatory* 125 (2005): 25–30.

Mufid, Abdul. "Unification of Global Hijrah Calendar in Indonesia: An Effort to Preserve the Maqasid Sunnah of the Prophet (Saw)." *Journal of Islamic Thought and Civilization* 10, no. 2 (2020): 18–36.

Mufid, Abdul, and Thomas Djamaluddin. "The Implementation of New Minister of Religion of Brunei, Indonesia, Malaysia, and Singapore Criteria Towards the Hijri Calendar Unification." *HTS Teologiese Studies / Theological Studies* 79, no. 1 (2023): 1–8. https://doi.org/10.4102/hts.v79i1.8774.

Nawawi, Mohd Saiful Anwar Mohd, Khairussaadah Wahid, Saadan Man, Nazhatulshima Ahmad, and Mohammaddin Abdul Nir. "Pemikiran Imam Taqī Al-Dīn Al-Subkī (683/1284-756/1355) Berkaitan Kriteria Kenampakan Anak Bulan." *Jurnal Syariah* 28, no. 1 (2020): 1–30.

Neugebauer, Paul Victor. *Astronomische Chronologie.* Berlin: Walter de Gruyter & Co., 1929.

Niri, Mohammaddin Abdul, Raihana Abdul Wahab, Mohd Saiful Anwar Mohd Nawawi, and Abdul Razak Nayan. "The Knowledge Integration Perspective on the Issue of Determining the Time for the Beginning of Fajr Prayer." *Jurnal Fiqh* 16, no. 2 (2019): 253–88.

Odeh, Mohammad. "New Criterion for Lunar Crescent Visibility." *Experimental Astronomy* 18 (2004): 39–64. https://doi.org/10.1007/s10686-005-9002-5.

Qureshi, Muhammad Shahid. "A New Criterion for Earliest Visibility of New Lunar Crescent." *Sindh University Research Journal (Sci. Ser.)* 42, no. 1 (2010): 1–16.

Ramadhan, T. B., Thomas Djamaluddin, and J. A. Utama. "Reevaluation of Hilaal Visibility Criteria in Indonesia by Using Indonesia and International Observational Data." Paper presented at the Proceeding of International Conference On Research, Implementation And Education of Mathematics and Sciences, Yogjakarta, 2014.

Rhodes, Brandon. "Skyfield: High Precision Research-Grade Positions for Planets and Earth Satellites Generator." Astrophysics Source Code Library, 2019. https://ascl.net/1907.024.

Schaefer, Bradley. "An Algorithm for Predicting the Visibility of the Lunar Crescent." *Bulletin of the American Astronomical Society* 19 (1987): 1042.

Schaefer, Bradley. "Astronomy and the Limits of Vision." *Vistas in Astronomy* 36 (1993): 311–61.

Schaefer, Bradley. "Atmospheric Extinction Effects on Stellar Alignments." *Journal for the History of Astronomy* 17, no. 10 (1986): S32–S42. https://doi.org/10.1177/002182868601701003.   http://journals.sagepub.com/doi/10.1177/002182868601701003.

Schaefer, Bradley. "Heliacal Rise Phenomena." *Journal for the History of Astronomy, Archaeoastronomy Supplement* 18, no. 11 (1987): S19–S33.

Schaefer, Bradley. "Length of the Lunar Crescent." *Quarterly Journal of Royal Astronomical Society* 29 (1991): 511–23.

Schaefer, Bradley. "Lunar Crescent Visibility." *Quarterly Journal of Royal Astronomical Society* 37 (1996): 759–68.

Schaefer, Bradley. "Lunar Visibility and the Crucifixion." *Quarterly Journal of Royal Astronomical Society* 31, (1990): 53–67.

Schaefer, Bradley. "New Methods and Techniques for Historical Astronomy and Archaeoastronomy." *Archaeoastronomy* 15 (2000): 121–25.

Schaefer, Bradley. "Telescopic Limiting Magnitudes." *Publications of the Astronomical Society of the Pacific* 102, no. February (1990): 212–29. https://doi.org/10.1086/132629.

Schaefer, Bradley. "Visibility Logarithm." In *Sky & Telescope*, 57–60. Cambridge: Sky Publishing, 1998.

Schaefer, Bradley. "Visibility of the Lunar Crescent." *Quarterly Journal of Royal Astronomical Society* 29 (1988): 511–23.

Schmidt, J. F. Julius. "Ueber Die Früheste Sichtbarkeit Der Mondsichel Am Abendhimmel." *Astronomische Nachrichten* (1868). https://doi.org/10.1002/asna.18680711303.

Schoch, Carl. "The Earliest Visible Phase of the Moon." *The Classical Quarterly* 15, no. 194 (1921): 3–4.

Siedentopf, H. "New Measurements on the Visual Contrast Threshold." *Astronomische Nachrichten* 271 (1940): 193–203.

Spain, Don. "Apennine Mountains." In *The Six-Inch Lunar Atlas*, 119–22. Springer, 2009.

Sultan, A. H. ""Best Time" for the First Visibility of the Lunar Crescent." *The Observatory* 126 (2006): 115–18.

Sultan, A. H. "First Visibility of the Lunar Crescent: Beyond Danjon's Limit." *The Observatory* 127 (2007): 53–59.

"The Future of World Religions: Population Growth Projections, 2010–2050." Pew Research Center, 2015. https://www.pewresearch.org/religion/2015/04/02/religious-projections-2010-2050/.

Vinayak, Mehta. "Camelot-Py." Python Package Index, 2023/2/. https://pypi.org/project/camelot-py/#description.

Yallop, B. D. *A Method for Predicting the First Sighting of the Crescent Moon*. Nautical Almanac Office. Cambridge: Nautical Almanac Office, 1998.

Yallop, B. D. "A Note on the Prediction of the Dates of First Visibility of the New Crescent Moon." In *Astronomical Information Sheet*, 1998.

Index

Printed in the United States
by Baker & Taylor Publisher Services